DATA STEWARDSHIP
FOR
OPEN SCIENCE

Implementing FAIR Principles

DATA STEWARDSHIP
FOR
OPEN SCIENCE

Implementing FAIR Principles

BAREND MONS

CRC Press
Taylor & Francis Group
Boca Raton   London   New York

CRC Press is an imprint of the
Taylor & Francis Group, an **informa** business

A CHAPMAN & HALL BOOK

CRC Press
Taylor & Francis Group
6000 Broken Sound Parkway NW, Suite 300
Boca Raton, FL 33487-2742

© 2018 by Taylor & Francis Group, LLC
CRC Press is an imprint of Taylor & Francis Group, an Informa business

No claim to original U.S. Government works

Printed on acid-free paper
Version Date: 20180216

International Standard Book Number-13: 978-1-4987-5317-3 (Paperback)
International Standard Book Number-13: 978-0-8153-4818-4 (Hardback)

Visit the Taylor & Francis Web site at
http://www.taylorandfrancis.com

and the CRC Press Web site at
http://www.crcpress.com

Printed and bound in the United States of America by
Edwards Brothers Malloy on sustainably sourced paper

To all Data Stewards and Eleni
To Rob, Robert, Marek and Joke, you know why...

Contents

List of Figures

Preface

When I was asked to write this book in early 2015, my first reaction was: "Why write a book with the essential message to stop writing?" However, after giving it some thought, and after talking to a few colleagues, I decided that this book may serve a purpose after all.

I fully understand that a printed book in (and about) the era of data-driven science seems like an anachronism. But, first of all, this book has an e-book version and, much more importantly, it is for a large part accessible in open access in the context of a data stewardship wizard. In the wizard there will be regular updates, as well as community participation in further improving the essentials for good data stewardship for open science.

Since the agreement to write this book, a lot happened. The FAIR principles were 'going viral', I was appointed chair of the High Level Expert Group of the European Commission to give advice on the European Open Science Cloud (EOSC), and meanwhile I co-led an international implementation strategy to realise the Internet of FAIR Data and Services. In all preparatory and advisory roles, I was confronted from all sides by the pressing need to educate and adequately equip a whole new generation of data stewards, who know how to deal with (FAIR) data as well as with the associated services. During the two-year period in which this book and the associated wizard developed, my motivation to contribute to the toolbox needed for data stewards only increased, and, here is the first result, for what it's worth. I hope it contributes to the development and establishment of a new, much needed, profession and ultimately to better science for a more effective society.

Author

From
http://www.marietmons.nl

Barend Mons is a molecular biologist by training (PhD, Leiden University, 1986) and spent over 15 years in malaria research. After that he gained two decades of experience in computer-assisted knowledge discovery, which is still his research focus at the Leiden University Medical Centre. He spent time with the European Commission (1993-1996) and with the Netherlands Organisation for Scientific Research (NWO 1996-1999). Dr. Mons also co-founded several spin-off companies in the field of semantic technologies and data curation.

Currently, Dr. Mons is professor of biosemantics in the Human Genetics department of Leiden University Medical Centre. He was the first head of node for ELIXIR-NL at the Dutch Techcentre for Life Sciences, he is integrator for life sciences at the Netherlands eScience Centre, and board member of the Leiden Centre of Data Science. He coined the Knowlet concept in 2008 and the concept of nanopublication, together with Jan Velterop, in 2009. Both are used in machine-assisted knowledge discovery. In 2014, Dr. Mons initiated the Lorentz Conference, which led to the FAIR data principles. In 2015, he was appointed chair of the European Commission's High Level Expert Group for the European Open Science Cloud (EOSC), DG Research and Innovation. He resigned from that group after completion of the EOSC recommendations and is currently leading the Global Open FAIR initiative.

Having been confronted on all levels by an alarming lack of awareness about good data stewardship, Dr. Mons wrote this book to make scientists, funders, and innovators in all disciplines and stages of their professional activities broadly aware of the need, the complexity, and the challenges associated with open science, modern science communication, and data stewardship. The FAIR principles will be guiding the reader throughout, and it should be emphasised here that these are indeed principles and not standards in and of themselves. Faithful to that, this book will not go into depth concerning specific standards and protocols, as good data stewardship and data can be realised in many different ways and will be different by discipline as well. Thus, this book should leave experimentalists consciously incompetent about data stewardship and hence motivated to respect data stewards as representatives of a new profession, while it might motivate others to actually become data stewards. However, just reading this book will not make you one.

CHAPTER 1

Introduction

1.1 DATA STEWARDSHIP FOR OPEN SCIENCE

Generating knowledge using public funding implies public responsibility.

Society and science are in an unprecedented transition, and those who deny the paradigm shift in science will soon be left behind. Science in the 21st century is no longer an ivory tower hobby of an eccentric elite, but it has invaded all sectors of society. Terms like *evidence-based* and *data-driven* emerge everywhere and suggest that properly collected and interpreted data and evidence is driving decisions and developments in almost every societal domain. Still, science should not be predominantly driven by potential practical applications. It serves the expansion of our collective knowledge and therefore it should by definition go in all directions. However, it is a logical consequence of the growing involvement of the general public in knowledge discovery that the research process is under more public scrutiny and pressure than ever before. Europe alone annually spends close to 250 billion Euro of taxpayer money on research and development, and the research and innovation process is increasingly assisted and scrutinised by well-educated citizens.

At the same time, and maybe related to this increased societal integration, visibility, and scrutiny, many elements of the scientific process itself that have been established for centuries are now facing a crisis. This includes peer review reproducibility, publication pressure-induced fraud, and undue ownership claims in scientific results that were generated with public funding. The credibility crisis we currently face in science is directly related to the fact that the way we value and communicate scientific output is stuck in the 20th century. The scholarly communication system and market did not yet pick up on the enormous power of the Internet and is dominated by an extremely profitable and hence conservative private sector. Before bashing the publishers for all this, which is too easy, we should realise that the scientific guild let the crucial business of communicating its results and re-using them for further knowledge discovery be orphaned, and thus handed it over to an industry that became so extremely profitable that it is almost unable to change without upsetting its shareholders. In the more recent past, some rescue operations have been undertaken, such as open access business models and the introduction of supplementary data to be co-published with articles, but these have not led to a fundamental rethinking and a step-change in the way we communicate science. They have led to more open access articles, which is a ministep forward, but also to the mushrooming of institutional repositories, in which data and text are open, but hardly findable, badly accessible, frequently non-interoperable, and thus still useless. As a result, preliminary studies indicate that in data-intensive science disciplines, the average PhD student may spend as much as 60% of research time on so-called data munging (trying to find, extract, reformat, and integrate data for meta-analysis). Better data stewardship should therefore not be seen as a new investment, but rather as part and parcel of the responsible use of public research funds and a major cost-saving factor, generating more capacity to actually perform creative research.

While we were sleeping, computers and in particular virtual machines rapidly became our most important research assistants, but we continue to make their jobs miserable by spitting out narrative, PDF's and other file formats that are near-useless for computers. More recently, we started to create rapidly decaying links to crucial supplementary data, rendering them either completely obscure, non-accessible, non-interoperable, or subjected to link rot and thus again reuseless, in particular for meta-research using machines. The scholarly communication and rewards system is thus trapped in the publish (narrative) or perish paradigm, and is no longer effectively supporting classical science, let alone open science. It will take a united effort, way beyond requesting open access articles, to break away from the 20th-century system that causes all the looming and substantiated crises in contemporary science. We cannot wait for governments, funders, or for that matter the status quo-oriented publishing industry to make that change, without the guilty guild taking the lead towards its own rescue.

As a first step towards this rescue operation for data and related services, we introduce here the recently developed but meanwhile widely adopted FAIR principles. Later on, these principles and their boundaries will be discussed in greater detail, but at this point it suffices to lay out the general principle that data (and the services to make sense of them) should be **F**indable (independently by machines and humans), **A**ccessible (under well-defined conditions), **I**nteroperable (again, independently by machines and human users), and thus **R**eusable (under properly defined licences and properly cited). In many cases data will be reused way beyond the purposes for which they were originally created, so long-term data stewardship will be a crucial element in open science. We define data and services that do not meet one or more of these elements of FAIRness here as reuseless.

The FAIR principles, as explained later in more detail, do not constitute a standard, nor do they specify a particular format or technology. Rather, they give a context and a direction to efforts to make data and services more useful and actually support their reuse. In that general sense, they are the major guiding principles for any form of proper data stewardship.

Throughout this book you will find a series of one-liners as take home (or Twitter) messages, which could be considered *virtual machine-outcries*. The first is:

Machine-readable research data are key and narrative should be supplementary.

Contemporary (open) science is increasingly based on reusing another's data and methods. As a key substrate for the computer-assisted knowledge discovery process, data should be machine-actionable wherever possible and be carefully stewarded for their entire life cycle. For the purpose of this book, we cover the entire process that deals responsibly with one's own and other people's data throughout and after the scientific discovery process under the term data stewardship. Taking good care of research data has obviously always been good scientific practice, but we all know that the current situation around research data is in total disarray. Now that science is in transition to a much more internationally collaborative and collective intelligence-based knowledge discovery effort using data from different disciplines, good data stewardship moves to centre stage. Although data stewardship itself may not be considered rocket science, good data and research infrastructure will be a prerequisite for future top-science and could thus be considered the rocket launcher.

Generating research data without an executable data stewardship plan is scientific malpractice.

So, what is open science, just a hype term or a fundamentally new way of doing science? The term open science is clearly already referring to different concepts in different people's minds and meanwhile enjoys many definitions. The most open societal definition of it might be found in the most prominent citizen-driven knowledge base, Wikipedia, which states: *"Open science is the movement to make scientific research, data, and dissemination accessible to all levels of an inquiring society, amateur or professional. It encompasses practices such as publishing open research, campaigning for open access, encouraging scientists to practice open notebook science, and generally making it easier to publish and communicate scientific knowledge"*. From a more methodological and computer science perspective, one could say that open science is an umbrella term for a technology and data driven systemic change in how researchers (and computers) work, collaborate, share ideas, disseminate and reuse results, by adopting the core values that knowledge should be optimally reusable, modifiable, and re-distributable.

The reuse aspect of open science is mostly undervalued in the most common definitions. Open science is so much more than open access of research articles. It is a new paradigm in the scientific method where meta-research over massive amounts of distributed data reveals myriads of patterns. A major new challenge in the data-rich era is to discern meaningful patterns and extract actionable knowledge from them. So, data publication, data stewardship, and reusable workflows to process these data in many different combinations are key prerequisites for open science.

Open science cannot develop without machine-readable data and good data stewardship.

Therefore, the printed version of this book serves only one purpose:

To make you fully aware of the importance of good data stewardship in open science.

In other words: this is mainly an **awareness-raising tool**, not a detailed methodological handbook with lists of detailed procedures. Actual procedures for good data stewardship frequently do not even exist yet, and will be continuously developed and described in online resources, protocols, and reference books. It has been a conscious choice of the author to keep the introductory part of this book short and hopefully valuable for many years in a field where new technologies, data types, analytical methods, standards, and best practices improve and develop on a weekly basis. The average life span of write and read technologies (remember VHS and the floppy disc?) and web services is on the order of years rather than decades.

Reading this book will make you aware of the complicated issues that make data stewardship a key professional skill in data-driven science, regardless of the phase or status of your scientific career. Consequently, training in data stewardship for years to come will have an important learning-by-doing element. Thus, this book and the associated online tool will be as much a learning tool as it is a broad practical guide to make conscious and responsible choices in data stewardship led by the now widely adopted FAIR principles.

Before reading this book, there are a few short animations to watch and a key paper to read that should convince you of the value of proper data stewardship for improved discovery.

1. "What Is Open Science All About, and how does it relate to data stewardship?"[1]

2. "Why Open Science Matters", a sci21 movie.[2]

3. "Why is Linked Open Data Critical?"[3]

4. "Why Are the FAIR Guiding Principles Crucial in Good Data Stewardship?"[4]

5. "What Are the FAIR Guiding Principles in Practice?" [Wilkinson et al., 2016]

---

[1] http://www.dtls.nl/5825-2/
[2] https://www.youtube.com/watch?v=7`9y3wbUgzU
[3] http://www.apple.com
[4] https://www.youtube.com/watch?v=PWutnWBfUSw

1.2 INTRODUCTION BY THE AUTHOR

WHY THIS BOOK?

Hi, I am an ageing molecular biologist who happens to find himself in the midst of the top-league of data scientists and computer specialists lately. I did not ask for that, my research ambitions led me into this cultural jungle. Many of these strange data-analytics people, while first appearing to speak a foreign language if not coming from a different planet altogether, now have become my friends and highly respected colleagues. Together, we have been able to do some amazing research that we would certainly not have been able to do (or even conceived of) if we had not developed what I consider to be the most important social-methodological skill of the modern scientist:

Open science needs respectful collaboration of specialists.

Still, I meet experimental researchers all the time (many of them much younger than I am) who are largely unaware of the enormous value of this collaboration between domain specialists (biologists, chemists, social scientists, you name them) and data specialists. Your data colleagues may frequently have little clue about what your data might actually mean, but they can magically understand the emerging properties and patterns from your data by combining them with hundreds of other datasets you were not even aware of and visualise them in graphical artwork that you would almost put on the wall in your living room. But what do these beautiful pictures actually tell you, or are they effectively ridiculograms (see Figure 1.1)?

That said, without your data specialist colleagues, you would probably still be ploughing through your spreadsheets, switching to another screen (if you know how to do that), and typing keywords in the literature or database search interfaces you happen to know. You probably miss more than 80% of what is out there in the Internet of Data (wish we had that?) and actually manage to republish in a so-called top scientific journal what was already explicitly provided as evidence in a database, unbeknownst to you and apparently also escaping the attention of your reviewers. This is not a joke, I come across these examples almost on a weekly basis in my current work.

However, what do these impressively visualised patterns mean, that magically emerge from your data, once they are combined with many other datasets and curated data resources through multiple computer

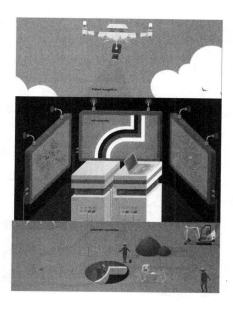

Figure 1.1   This picture is composed of several screen shots from the open science animation mentioned before. Two levels should be clearly distinguished in open science: Pattern recognition in big and sometimes noisy data requires very different technologies and methods as compared to fine-grained search for mechanistic explanations and causal relationships. The former process is compared to a drone spotting patterns on the ground that remain just patterns (or ridiculograms) until a digging phase has revealed the actionable knowledge derived from the patterns.

algorithms, the names of which you cannot even pronounce, and where a new one is coined as the best thing since sliced bread every month? *No idea*; that is where your domain expertise comes to the table. Your data specialist colleagues would obviously not start reading deep biology, chemistry, and sociological textbooks to become half-baked domain experts, would they?

SO WHY WOULD YOU READ BOOKS LIKE THIS ONE?

To become a half-baked data expert? Well, you should not. As said, this book is not meant to make you a half-baked data expert at all. It will not teach you much of the basics of data science or computer science (my computer and data scientist colleagues would have a good laugh if I tried). Instead, this book is aimed at the middle ground between the domain expert and the data expert. Any domain specialist in open, data-driven science should pay due respect to, and work closely with, data experts. This is far from trivial and a frequent reason for failure of projects, or even entire e-infrastructures (see Figure 1.2).

Figure 1.2 A prerequisite to conduct open and data intensive science is the respectful collaboration of domain specialists with data and computer scientists. However, these two expert communities do not necessarily communicate easily and are driven by very different reward systems and incentives.

Other people's data and services (OPEDAS): Whether you are an experimental scientist trying to interpret the data you just generated or a researcher in the humanities using data from social media, nowadays you will find yourself quickly using other people's data and, if you are clever, other people's analytical services. In fact you will more often than not use other people's data as well as other people's software, tools and computers.

The famous statement that every scientist stands on the shoulders of giants [5], is rapidly gaining a new connotation. Traditionally it mainly meant that we base our insights on knowledge we derived from prior scholarly communications (mostly talks at conferences, narrative, tables, and figures) but now it is applicable to founding our new discoveries on other people's prior research objects [Bechhofer et al., 2010]. As stated earlier, let's loosely define them as: any artefact produced or used in the scientific process. So, research objects comprise indeed data, algorithms, workflows, code, slides, video instructions, websites, tools, reference datasets, annotations, curated databases, and, yes, *even* narrative, and tables and figures. Here we will refer to this container concept as other people's data and services: OPEDAS.

A key feature of proper science is the reproducibility of studies, results, and conclusions. There are many reasons why study results may not be reproducible entirely and exactly the same in different experimental or social settings. Not being trivially reproducible does not automatically mean the conclusions drawn by the original conductors of the study were wrong. However, optimal care should be taken to make results as reproducible as possible, and good data stewardship is the very basis of such good research practice. In a recent American study [Freedman et al., 2015], it was shown that 25% of the non-reproducibility problem in pre-clinical research was related directly to problems in data analysis and reporting.(see Figure 1.3).

We will explain later how the other methodological aspects, reference materials, study design, and laboratory protocols are all of concern to good data stewards. That is why data stewards (core data professionals) should be embedded in every modern research team. Not only could the 28 billion dollar mentioned largely be saved, but also some lives while we are at it. The loss of valuable data, combined with an exploding ability to generate data, discover patterns in them and produce millions of correlations, has also contributed to the reproducibility crisis. A rigorous quality check on the supporting data for any conclusion in narrative is badly needed, and in many cases very difficult or even

[5]Mostly attributed to a 1676 letter of Isaac Newton, because it was in English but actually traceable back to the 12th century Bernard of Chartres

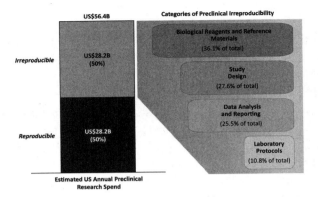

Figure 1.3 Data stewardship problems today and the major reasons for non-reproducibility, explained in the figure.

impossible, if only due to the sheer size of the data. Still, science should not become like a *fruit-machine* of spurious correlations based on data that are either lost, not accessible, or not reusable for replication or for inclusion in new studies (Figure 1.4). The lack of possibilities to rigorously check the sources and provenance of data has also led to some very visible cases of fraud, damaging the image of science overall.

We must conclude from the above that one of the most disturbing phenomena of the early 21st century is that while science transitions to Science 2.0 or open science, where research objects and machines that can operate them are increasingly used in many combinations to study and discover more and more complex associations and interactions, the scientific award and communication system is stuck in the 20th century and based on archaic and often misleading measures such as the journal impact factor[6]. This keeps narrative articles (designed exclusively for human reading) centre stage and this situation severely hampers the development of the data-driven science indexdata!driven science method. The guilty guild are not only the funders or the publishers, but also the scientific community who keep each other in this deadly *ego-embrace*.

From incidental surveys conducted under young PhD students, the

---

[6]I would hope you will soon need a footnote here to remember what that historical concept was.

Figure 1.4 Once big data is generated, it is easy to find many patterns and correlations in the datasets. However, a major skill of modern data stewards will be to help the domain specialists in discerning meaningful patterns, true correlations, and more importantly, to dig for the mechanistic explanations and causal relationships that lead to actionable knowledge. Access and reproducibility options regarding the original data presented as evidence for scientific claims is of the utmost importance to prevent non-reproducible results, sub-optimal conclusions, and outright fraud.

gloomy picture emerges that they have to spend roughly 50% to 70% of their time on a process that is now called *data munging or data wrangling*[7]. This means that before they can start their analytical process they need to extract, transform, and load (ETL) data from all kinds of sources and formats, and more often than not they are creating customised and amateur-hacked workflows to do the analyses,

[7]https://en.wikipedia.org/wiki/Data_wrangling

just because they do not know that the appropriate workflow already exists and may be re-used for their purpose. That said, many workflows that worked for the person who made them and are published in software repositories are indeed not findable, cannot be accessed, do not work any more, and can thus effectively not be re-used in actual practice. In a large academic hospital such as the one with which I am affiliated, the capital loss of over 800 PhD students at any given time may run into staggering figures of over 20 million Euro annually. So, the estimates that globally, we collectively lose many billions of taxpayer's money every year, just as a result of malpractices around data, seem very realistic to me. This range of 50 to 70 % of the time being spent on data wrangling in studies with a data integration component is found in multiple settings. Figure 1.5 (courtesy of Isabel Fortier) shows the same pattern. The example comes from a Swedish project where 21 cohorts of people had to be combined for studying 20 core variables. The research proposal was written in a matter of weeks, the actual identification of the cohorts took 3 years (2011 to 1014), the ethics application and approval took 6 months. These are largely unavoidable groundwork issues, although the identification of the desired cohorts to be included could also be significantly more efficient if all had FAIR metadata. However, during the project itself (running from 2012 to 2014), the actual data transfer from the cohort studies to a common analytics environment took 24 months, the data harmonisation another 18 months (partly overlapping), and finally, the actual pattern recognition and statistical analyses took 18 months, while the final manuscript was written again in a matter of weeks.

The cartoon drawn to try and visualise the (likely typical) labour-pattern in the project is shown in Figure 1.5. Although this is just an example, it is a very instructive educational cartoon. The majority of the foundational ground work in a data intensive project like this one could have been avoided if all cohorts had FAIR metadata and the data had been FAIR and harmonised *a priori*. The authors of one other exemplar data intensive study. They re-analysed 77,840 expression profiles and observed a limited set of 'transcriptional components' that describe and explain well-known cancer biology [Fehrmann et al., 2015], upon personal request and have reported that up to 85% of their time in this project needed to create the results was spent on data munging before the actual data analytics process could begin. I strongly believe that this undesired balance between data wrangling and actual data analysis sounds very familiar to all researchers dealing with data-intensive projects and using OPEDAS.

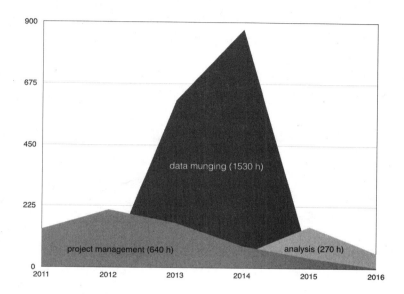

Figure 1.5 Data munging takes a very significant amount of time, especially in data intensive projects, and can be a major opportunity for increased efficiency and savings in scientific research and development.

Now that we want to enable the machine to be a first class research assistant, narrative journals, with machine-unfriendly ambiguities in language, tables, figures, and rapidly decaying links to supplementary data and a plethora of disconnected local data repositories, are about the worst way imaginable to communicate the outcome of the scientific process. If science has become indeed data driven and *data is the oil of the 21st century*, we better put data centre stage and publish data as first-class research objects, obviously with supplementary narrative where needed, steward them throughout their life cycle, and make them available in easily reusable format.

Yet another recent study claimed that only about 12% of NIH funded data finds its way to a trusted and findable repository. Philip Bourne, when associate director for data science at the U.S.A. National Institutes of Health coined the term *dark data* for the 88% that is lost in amateur repositories or on laptops. When we combine the results of the general reproducibility related papers and the findability studies, the picture is pretty gloomy. In a set of 3.5 million articles ranging from 1997 to 2012, the Hiberlink team found that in articles from 2012, 13% of hyper-links in arXiv papers and 22% of hyper links in papers from Elsevier journals were rotten (the proportion rises in older articles), and overall some 75% of links were not cached on any Internet archiving site within two weeks of the article's publication date, meaning their content might no longer reflect the citing author's original intent, although the reader may not know this (see Figure 1.6).

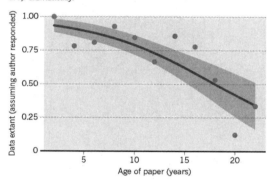

MISSING DATA
As research articles age, the odds of their raw data being extant drop dramatically.

Figure 1.6 Link rot. Missing data;:The links to supplementary data files in narrative papers are subject to significant decay over time. This results in poor findability of these data, especially for machines. See original reference figure in Vines, T. H. et al. Curr. Biol. $http: //dx.doi.org/10.1016/j.cub.2013.11.014$ (2013)

THE BANKRUPTCY OF THE ARTICLE[+] APPROACH

In open science , open access to articles is obviously desirable. However, it is a major misconception to assume that open access articles will automatically support and empower the process of open science. Data-driven science in particular can still be enormously hampered by the current *Article[+] approach* in scholarly communication.

What is the Article[+] approach? Let's define it as the old-fashioned scholarly communication practice upholding the tenet that the principle unit of scientific communication is the textual article with some stuff added to it (+). Figures and tables are typically still part of the paper, but supplementary data are frequently linked to it and stored in a random separate location. It does not require a lot of imagination to picture how we thus create an outright nightmare for machines. Already, for example, in text-mining, the problem is imminent. Here we distinguish 3 subsequent firewalls for machines.

The first firewall occurs obviously when virtual machines (for instance text and data-mining workflows) are simply denied access to the text. Social change and pressure to open up this first line of defence for publishers is building rapidly, and systems are also put in place to mine individual assertions from restricted full text, so this first firewall is going the same way as the Berlin wall and will crumble soon, although many years too late. As a consequence, publishers are increasingly recognising the inevitable, and have started to expose the key assertions from the full text of their closed articles in machine-readable format and also increasingly in open access, and/or they allow text-mining on their corpus of text. After all, they see that papers are meant for people and in the end people will want to come and do confirmational reading on the full text, even if they have to pay. Exposing individual assertions and claims from their proprietary text in open science analytics tools and environments can make their individual and isolated assertions function in pattern recognition, while luring people to the original pay walled paper for conformational reading. So for the purpose of this book, we consider the pay wall first line of defence crumbling. This obviously does not mean that open access pressure should go off. However, we should also not see open access text as a magic solution, nor is open access morally very different from the old model. In the old model, underprivileged colleagues (for instance, in low-income countries) could not easily *read*. In the open access model the publication fees are frequently too high for them, so they cannot *publish*, putting them once again in a disadvantaged position. Furthermore, even if machines get access to text, they can only mine meaningful assertions and information from it with sophisticated text-mining and disambiguation algorithms. In fact, a whole scientific sub-discipline has developed with the sole aim of rescuing information buried in narrative and mining it for machine-actionable information [Mons, 2005].

The second firewall is the machine-unfriendly format of tables and figures in articles and the combination with free text. If, for instance, the text says: all countries in Table 1 drive on the left, it is already quite a challenge for a machine to make sense of that, even if the table can be text-mined and the thesaurus contains all the countries involved. If, however, the table is in a file format that escapes text-mining altogether, as is common practice, the problem is aggravated [Rebholz-Schuhmann et al., 2012].

The third firewall, and probably one of the most deleterious ones, is the wall of broken links. A hyper-link to supplementary data as such does not mean much to a machine, and if the metadata of the supplementary files are not FAIR and properly exposed in the publication, much of the supplementary data will be effectively hidden behind the article and missed by machines. If the link is dead anyway, the machine will be stuck altogether, and even if an algorithm was clever enough to follow the link and attempt to mine the supplementary data, the actual data may also appear more often than not in a machine-unfriendly format [Mons et al., 2011].

Two instructive examples from the life sciences

The first instructive example is described in a paper by Yepes and Verspoor (2013) [Jimeno Yepes and Verspoor, 2014]. It appears that the vast majority of single-nucleotide polymorphisms (SNP, mutations) in the genome that have been recorded in a plethora of dispersed databases around the world is missed if one tries to text-mine them back from MedLine, the major reference database in the biomedical life sciences. I have tried to summarise the major points of the paper, combined with common knowledge about the general subject area for educational purposes (Figure 1.7).

The first row depicts that the average abstract (abstracts are usually open access) may contain about five meaningful assertions. Probably more, but that seems to be the high end of what we are currently able to reconstruct by text-mining. The average full-text article (the older ones mostly still behind pay walls) will maybe contain in the hundreds of assertions. Tables and figures (behind the TIF firewall, even if all articles would be open access) may be in the range of hundreds to thousands of associations, and some (supplementary, underpinning) data may contain many more (up to millions) of associations. They will usually be linked as supplementary data. The row below shows you just how inadequate this Article(+) approach is to scholarly communication nowadays: single-nucleotide polymorphism (SNP)-Phenotype associations that are known from all collective databases cannot effectively be recovered/reconstructed from MedLine abstracts. In fact, the performance is not much better than 2% (!) of the total. Maybe surprising to you at first glance, adding full text to that (assuming the first fire

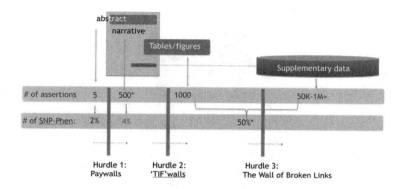

Figure 1.7 The sequential hurdles put up by the $Article^{(+)}$ approach

wall has really fallen) would increase that figure only to a very disappointing 4%. Even if every wall is broken brick-by-brick by hand, we might still not be able to recover much more than 50% of all this crucial information from the current, broken $Article^{(+)}$ approach. Obviously, open access $Article^{(+)}$ approaches alone would not significantly improve the broken system from this perspective. In any case, there is far too much for people to read, and MedLine currently grows with one article abstract every 40 seconds. If a person would want to read everything about a common disease like diabetes it would take about 68 years if one could keep up reading and conceptualising 20 scientific articles per day (personal communication Isabel Fortier, 2016, and see Figure 1.8).

Still, people will always want to read articles and follow other people's extensive reasoning. Apart from the fact that there is too much to read for the human brain in most domains, it is however of the utmost importance to publish data and the main claims in narrative first and foremost in machine-readable format. We have reached the point where integrating all published data and curated information leads to a display of such complexity that the human mind is unable to discern meaningful patters in the resulting ridiculograms. A frequent reaction is that people turn away from the apparent complexity shown by the

Figure 1.8 No escape from writing more than we can read? *Article*$^{(+)}$ *approach*

data and turn back to reading and attempts to distil the information into subsets that are intellectually manageable (see Figure 1.9). However, the complexity in many hairball type meta analysis outputs may very well be extremely relevant. A major challenge, and certainly in the realm of good data stewards, is therefore to *re-fragment* the complexity of massive data analytics into meaningful components that are understandable for humans, so that a social machine approach becomes feasible.

In conclusion to this paragraph, my statement in 2005: Text-mining? Why bury it first and then mine it again? [Mons, 2005] is still frighteningly relevant.

Figure 1.9 No escape from complexity $Article^{(+)}$ approach

A good data steward publishes data with a supplementary article(Data(+)).

One more example, that also demonstrates how the system can be changed without completely breaking it and turning away from the old system is illustrated very nicely in the FANTOM5 case. In FANTOM5, a large group of research institutes, led and coordinated by the RIKEN Institute in Japan, have systematically investigated the sets of genes used in virtually all cell types across the human body, and the genomic regions that determine from which points the genes are transcribed. The consortium has mapped the sets of transcripts, transcription factors, promoters, and enhancers active in the majority of mammalian primary cell types, and a series of cancer cell lines, and tissues, which is described in the landmark paper in Nature. [The FANTOM Consortium and the RIKEN PMI and CLST (dgt), 2014]. Around 30 publications derived from the project cover areas as diverse as primary cells, gene families, genome wide observations on promoter features, and new bioinformatics tools. The mothership paper, describing the study that generated the dataset itself has over 350 authors from more than 95 different institutions, and 24 additional files. Obviously, the mothership paper can only describe (in text) a minimal part of what can be derived from the dataset that describes how more than 90,000 genomic entities co-regulate the expression of about 20,000 genes in all tissues. The real data are referred to in the paper as: All CAGE data have been deposited at DDBJ DRA under accession number DRA000991. There is supplementary information (5.6 MB) in PDF, and in Excel (36.8 MB). You figure out how machines can activate this data if it is not FAIR. Our group at LUMC participated in the publishing process by constructing so-called nanopublications [Groth et al., 2010] as supplementary material. If we look at only three types of nanopublications from FANTOM5 data: CAGE peaks; their associated genes (type II), and their expression information (type III), we already count 51,942,135 nanopublications, representing only a minor fraction of the individually meaningful assertions that came out of the study. So, although I would consider the FANTOM5 consortium among the most advanced groups in open, collaborative science, making this paper open access does not solve the problems computers would have to independently find, access, link, and reuse the enormously important information created in the FANTOM5 study. Modern data stewardship should allow automated queries over machine -readable and citable data of the type shown in Figure 1.10 (courtesy of Rajaram Kaliyaperumal, LUMC).

Figure ?? indicates how a computer would be able to independently reason that the deletion found in a locus on chromosome 2 disrupts a putative transcription start site as found in FANTOM5, and may therefore influence the expression of the gene under its control and cause a disease, as recorded in the independent database LOVD [8]. This is what open science should be all about: machines and people can seamlessly interact as *social machines* to perform automated and verified knowledge discovery in concert. For that to happen, much more than open access publications is needed, a cultural change first and

---

[8]http://www.lovd.nl/3.0/home

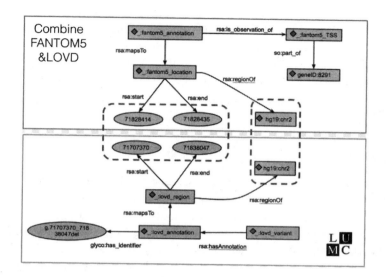

Figure 1.10 Variants in the Leiden Open Variation Database connected to TSS candidates in the Fantom5 study via linked data as an example of the possibility to use distributed and independently generated linked FAIR data to discover new knowledge.

foremost, but also new ways of scholarly communication and rewards and many other issues, with at its core the Data(+) approach.

At the very basis of the needed revolution in scientific methods lies good data stewardship.

For the transition to open, data-driven science to be smooth and efficient, many more socio-cultural hurdles have to be taken than technical ones. For most of the technical challenges facing us in data-intensive science, experts find incrementally better solutions pretty fast. However, as these efforts are not properly awarded and supported, the experts who can deal with data are undervalued, do struggle with job security, and frequently disappear to better paid jobs, leaving academic science crippled and stuck with half-baked solutions. We will not be able to fully exploit the enormous value of data unless we fundamentally change our attitude and learn from disciplines like high-energy physics or astronomy, where scientific and data engineers have had a proper status for decades, Therefore, regardless of whether you are a senior scientist, or even a science manager approaching retirement, or an ambitious young student gearing up to spend a life in science, you will need to be aware of the enormously complex issues associated with optimal reuse of other people's stuff (OPEDAS). And you will be taking part in a methodological landslide whether you like it or not, unless you want your research institution to be left behind in the reading and writing age.

HOW DID WE END UP IN THIS MESS ?

For centuries, researchers in many scientific disciplines have done experiments and created research data. When systematic, empirical science emerged in the 16th and 17th centuries, the focus on supporting data gradually increased, but the scientific method was still dominantly hypothesis-driven. When I was trained as a cell biologist (late 1970s) we were told to *read everything on the subject* before starting an experiment. Sit back and reflect for a moment on what such a statement would mean today, only 35 years later.

How science will be conducted another few decades ahead, we simply cannot predict. A few things are, however, very obvious, although they seem to escape many scientists today. First, our technical ability to generate data, both for research and in society at large, far out paces our abilities to make optimal use of those data for knowledge discovery and innovation. The statement that 90% of the total global data has been generated in the last two years[9] will possibly stay true for many years to come. Second, we will reveal levels of complexity in biological and other systems not even imaginable before the data started to expose the emerging patterns to us. Scientists, institutions, and possibly entire disciplines that ignore or undervalue the necessary adaptations of the scientific (communication) method and research infrastructure to the exploding possibilities data offer us will miss enormous opportunities and lose ground. Those who keep publishing their data (if they publish them at all) behind successive barriers for computers are wasting public money and should be punished for that in the new reward systems we need for data-driven science (see Figure 1.11).

It appears that science, although meant to propel innovation, is in itself a very conservative sector of society, functioning as an ego-system and carefully protecting its own self-imposed by-laws (see

---

[9]http://www.sciencedaily.com/releases/2013/05/130522085217.htm

Figure 1.11 Trains (a metaphor for contained virtual machines) that try to get to data in the current publishing ecosystem first encounter text. Text is a nightmare for computers, regardless of whether it is open access or not. That is why in this cartoon the first hurdle is deliberately not the pay-wall. Both for open and restricted texts the same non-machine readability applies. The next wall is the 'un-minable insert' wall discussed earlier. Even if the computer would be able to master all this, the links to the actual computable data are unintelligible links to web resources of which an increasing percentage is broken in the first place. This is why scientists need to spend the larger part of their PhD on data recovery, ETL and munging.

Figure 1.12). The Article (+) approach, including its unwanted side effects such as judgement of scientific status based on journal impact factor or H-factor type measures, may have been functioning pretty well in a data-sparse era. However, now that we are struggling not to drown in our own data, this scholarly communication approach is no longer in sync with the way we conduct our knowledge discovery and fails to offer the space to clearly communicate the complexity of the phenomena the data reveal. Stubbornly stating that only peer reviewed literature and curated databases are credible sources of scientific legacy information and hence scientists should (only) be judged on their measurable contributions to those sources, is a very unwise and actually dangerous attitude. Obviously, both of these sources will continue to be important elements of the scientific substrate. But data and services for knowledge discovery (OPEDAS) will become minimally as important. Creating those, as well as offering them in reusable format to other scientists, to innovators, and increasingly also to citizens, should be treated and rewarded as a major output of the scientific process. Consequently, research teams that produce data and services that can be reused by many others should be rewarded for that by the funding and tenuring systems. This critically includes professional data stewards in each and every scientific team or consortium.

Data stewardship is a key element of open science

Okay, let's stop complaining and see how we can solve the problem. Data management is a concept that has already been in use for a longer period of time. Most funders and researchers interpret it as taking good care of data during the research process. Obviously, this is very important, but in the era of data-driven and open science, many datasets have a useful life way beyond the project that originally generated them. Therefore, data stewardship, here roughly defined as: treating data and the associated research objects with the utmost care, with the aim to make them reusable for discovery as long as they are valid, is not a boring, unavoidable task for dusty, conservative people, but a rapidly developing profession at the very heart of modern science.

Before we go on, let me try to settle once and for all the the answer to the most frequent accusation I have endured for a decade from mostly very established scientists that made their career (like myself) largely in the age of narrative, tables, figures, and human-readable databases. I am NOT advocating to replace narrative scholarly communication entirely by machine-readable and -actionable communication formats. Simplistic ontologically coupled assertions of the type

Figure 1.12 Established scientists, who became so established in the narrative phase of scholarly communication are very used to judge the impact of science largely on one type of metrics, reflecting in essence: *How many people cite my papers.* First of all, we know that citation of your papers does not necessarily correlate well with the scientific or societal innovation impact of your research, but more importantly, these classical methods do ignore the value of other research objects that are reused by others. These include apps, work flows, visualisation methods, curated databases, and high-performance analytics environments, just to name a few. This metric jeopardises the career opportunities of young scientists in general and that of data stewards in particular.

[subject][predicate][object], even with very rich context and provenance (i.e. nanopublications), are not likely to replace rhetoric, argumentation in spoken or written human language or human-readable outputs in tables and figures any time soon. After all, why would I write a book to argue that we should stop writing?

Both human-readable outputs and machine-readable outputs have their rightful place in modern scholarly communication.

The most frequent mistake made in scientific communication and tool building today, however, is the conflation of services and tools made for pattern recognition in data and for intellectual confirmation. Much more about that later, but for now: Data and any other research objects should always be adorned with three things: (1) a unique and president identifier, (2) rich, machine-readable metadata/provenance AND (3) a narrative, human-readable supplementary article describing the essentials to be known and argued about the genesis and the first interpretation of the data. Obviously, data stewardship entails a lot more than that, and this book will try to give you a broad overview of many questions you will have to ask yourself and colleagues in dealing with complex data.

The focus and limitations of this book

This book is not just for data stewards, it is certainly also not aimed at data scientists *per se*. It is also not a purely technical handbook with many standards and procedures to follow. In fact , data stewards will need other books and the links provided in the open-access data stewardship wizard interactive with the open access part of this book to study and develop the detailed skills of their trade. It is also not just for young humanities, biology, chemistry, geology, or physics students and also not just for senior principal investigators and managers caught in the landslide.

It is meant to be a general introduction to the various aspects of data stewardship in the data life cycle and thus useful to study for all professionals involved in today's complex and multidisciplinary science process. So, are you (or do you want to become) an experimental scientist, a data scientist, a data steward, a citizen scientist, a medical doctor, a business person, a funder, or any other decision-maker relying on data-supported and solid evidence? As said, you need to be aware of the many pitfalls and challenges around complex research data. That does not mean you have to be an expert in every aspect of the data, tools, and discovery cycles needed for your research. This book is meant to teach you awareness of the challenges in your day job, and respect for those who have dedicated their professional life to aspects you are consciously incompetent about after reading this book.

We can all agree on one basic principle: If whatever you do is based on suboptimal (or worse, the wrong), data, your next steps in the discovery or decision-making process will likely take you in the wrong direction. So, don't think that this book does not concern you and that it is only for data nerds. As argued earlier, in modern (e)science your substrata for knowledge discovery and decision-making will increasingly be already-existing data and services (OPEDAS) combined with your own data. For the proper use of this goldmine, you need professional data stewards next to you in every single step of your scientific activities and in your future decision making processes. (and although they frequently are, *never call them nerds again*). Reciprocally, for data professionals, if the next pattern you discern in massive, functionally interlinked data excites you even though it might just be an anomaly in your ridiculogram, you need domain experts next to you in every single step of the translation in order to transform patterns into actionable knowledge.

Trying to be a domain expert as well as a data expert will make you mediocre in both.

This book will therefore not lead you into deep technological discourse, standards, code, or scientific methods. It will cover the basics of all aspects of good data stewardship, but the actual text is meant to be understandable for all categories of interested students and professionals mentioned above.

Most experimental scientists (*in spe*) will conclude after reading this book that they have to critically depend on core data professionals in every step of their experimental and analytical process, while data stewards will be able to get in contact, via the associated wizard, with the best of the best in the respective technical communities they need to rely on for their profession. All other readers will be fully aware of all the issues influencing the data on which they may base their decisions, and hopefully be much more aware of the many confounding issues around data that will heavily affect the reusability of data and the conclusions on which their decisions may be based.

THE FORMAT AND INTENDED USE OF THIS BOOK

So how are we going to lead you through this process in a logical, and not -too- boring journey? The book will be built around the Data (Stewardship) Cycle as depicted in Figure 1.13.

After a general introduction, each of the following chapters will cover one of the boxes in the schematic research data cycle depicted in Figure 1.13. Experimental design, data design, and planning, followed

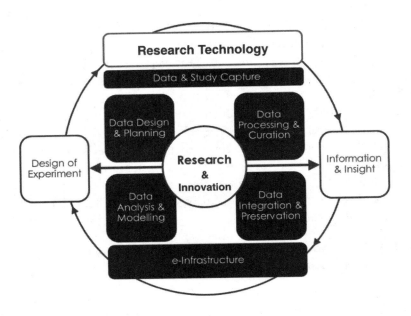

Figure 1.13 The data stewardship cycle:Data stewards should be involved in all phases of the research life cycle. They should sit on the team that designs and plans data-intensive experiments, they should co-supervise data capture, be involved in the data processing and curation, followed by linking, integration, and preservation processes, plan for the required e-infrastructure to run the experiments, the data analytics, but also to preserve the data and the planning and budgeting for offering the data for reuse. Finally, although this phase will also include hard-core data scientists and statisticians, the data steward should also have good knowledge about available methods and tools for data modelling and interpretation, which in turn may lead to new experiments to continue the cycle.

by the actual data capture as well as capture of data about the study. The next chapter deals with data processing and curation issues, with a special emphasis on preparing data for integration (or in actual fact, functional interlinking with other data) and preservation. There will be sections about the needs and choices around e-infrastructure needed to process, store and analyse the data, but no details on hardware

and connectivity itself. Finally, data stewardship issues pertaining to
the (linked and distributed) analysis of data will be addressed, but
again, this book is not a data analysis protocol handbook. There is no
shortage of books and literature about hard-core data and computer
science, hardware and analytical software programmes. These aspects
will therefore only be touched on lightly in this book. However, for data
stewards, it should be a guide towards proper choices and best practices
and therefore, at the end of each section, there will be a list of **DOs**
and **DONT'S**, followed by a link to a page with external resources,
which is jointly kept updated by the community contributing to the
associated data stewardship wizard, and thus will provide a growing
list of links to the most updated relevant web resources dealing with
the topic covered in that section. There, the standards, best practices,
active communities, and running projects to connect to will be listed
for further exploration and consultation.

It will be immediately obvious to you that such a set up is at
odds with paper print. That is why the colour version of this book is
only available as an e-book. When you access the e-book that comes
with this printed version, you will be able to both read and contribute
in the associated open access data stewardship wizard environment[10],
with the special purpose to always provide the very latest version of
updated and active links to the web resources you will need to take
action. So, the e-book is where regular updates will be printed and the
wizard is where the most recent links and activities can be found!

The current wizard is developed in the context of several national
nodes in the ELIXIR network, so again, it may be seen as specific for
the life sciences, but the set up will really be as depicted in Figure
1.14, with a central core on domain overarching issues and specialised
sections for domain specific issues. Figure 1.15 provides a screen shot
of version 1.0 (beta) of the wizard (courtesy of Elixir Czech Republic
and Elixir Netherlands).

1.3 DEFINITIONS AND CONTEXT

Before we can go any deeper into the matter, we need to further define concept-denoting terms that we will use a lot, such as data, digital objects, research data, research objects, FAIR principles, concepts,
modern science, open science, and data stewardship.

---

[10]http://dmp.data.solutions/

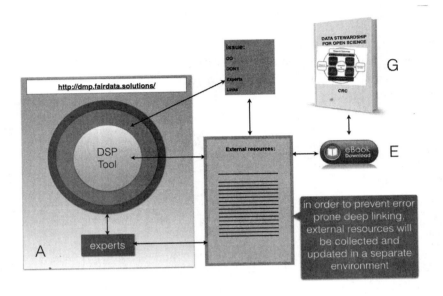

Figure 1.14 The text of each practical page of this book is linked to a corresponding page in a continuously updated data stewardship wizard. The online pages in both the e-Book and the Wizard lead to an 'external resources' page, which is kept up to date by the data stewardship community and provides back ground reading material, stewarded links to web resources, standards, best practices etc. These pages can be adapted per domain and linked towards in localized versions of the wizard. In localized wizards, local experts can be added to the pages.

DATA

As *data* and how we deal with them in science and innovation is the core subject of this book, it is worthwhile to spend some more time on defining exactly what we mean by the general term *data* and the various specifications of it.

First of all, the term data is used in many different ways in many contexts, for any set of measured values of qualitative or quantitative variables, regardless whether these are recorded on paper or in digital format. Many datasets may not be used for research, at least initially. In this book we specifically deal with *research* data, regardless of whether

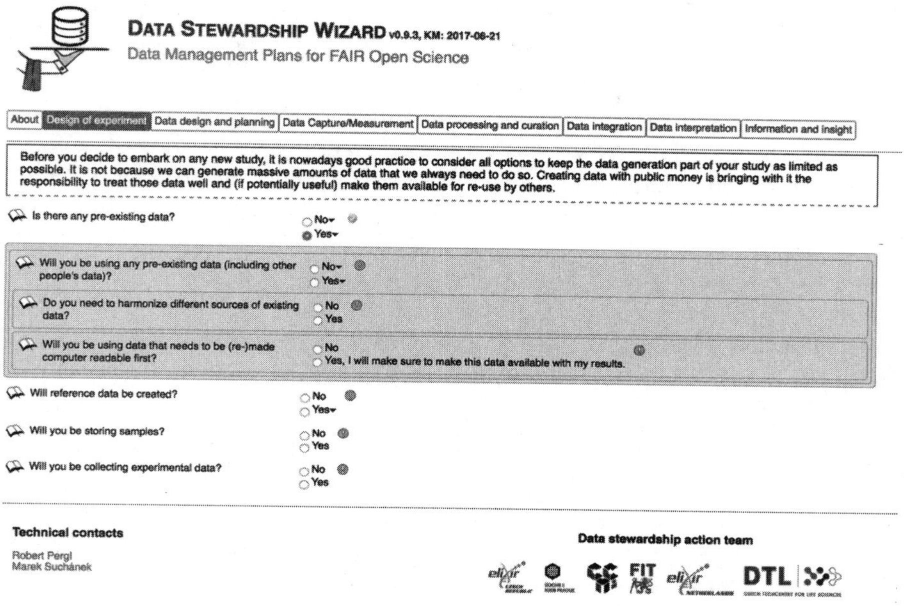

Figure 1.15 Screen shot of a data stewardship wizard as produced by the Czech and Dutch Elixir nodes. The wizard leads the data stewards through a large number of questions and assists in making a proper data stewardship plan. When the registered user clicks on the book icons on the left hand side of each question covered in this book, the corresponding book page will pop up with a link to external resources (see fig. 1.14). The links at the bottom of each practical page in this book will lead to the same external resources page.

it was purposely generated for research (i.e., in experimental settings) or whether the data just happens to be used for research later, for instance, data gathered in social media settings.

As we have irreversibly entered the digital age, also in science, we do not talk here about data in paper notebooks. We talk about data that is digital, in the sense that it was measured, collected and reported in digital format and is thus already in a form that they can be analysed, visualised, and interpreted in principle by machines. According to the general Wikipedia article on this subject, *raw data*, i.e., unprocessed data, is *a collection of numbers and/or characters*. In many cases, before the processes of analysis, interpretation, sharing and reuse start, we first process the raw data. Already here, confusion may come in, because data processing occurs in multiple stages and in many different ways in different disciplines, so that some people may feel that they do analysis directly on *raw* data. In fact, the processed data from one stage may be considered the raw data of the next. Stewardship issues are relevant for all levels, from raw data to processed data to metadata to turning data into information, and finally even knowledge.

DATA AS DIGITAL OBJECTS

In fact, anything, however small or large, that dwells as a bit string in repositories or on the Internet, can be seen as a *digital entity* (or a *digital object*). A digital object/entity is defined by the U.S.A.-based Corporation for National Research Initiatives (CNRI)[11] as: *An entity or object represented as, or converted to, a machine-independent data structure consisting of one or more elements in digital form that can be parsed by different information systems.*

RESEARCH DATA

The simplest specification of research data may seem that all data that are either produced or used in research fall under this specification. This means that data that were not originally created or collected with a research purpose in mind may become research data later, because they might reveal important patterns and information of relevance to a given hypothesis to be tested.

Most data will be used for research and decision making sooner or later, regardless of whether the data was originally created with research purposes in mind. In data-driven science, which is much less hypothesis-driven than classical science, unexpected combinations of data and resulting emerging properties in such combined datasets actually drive major discoveries, which is likely to be of particular importance in cross-disciplinary studies.

Data are crucial for many processes *successive* to scientific discovery: Reproducibility checks, innovation, decision-making, and many other aspects of society rely heavily on data. Thus, the premise of this book is that all decisions in society should ideally be based on

---

[11]https://www.cnri.reston.va.us

solid and comprehensive knowledge, and that all knowledge is sooner or later created by a form of scientific discovery process. Even when we think we make intuitive decisions, more often than not these will be based on prior experience and knowledge, whether we are conscious about that or not. So, let's agree that when we say research data in this book, we effectively mean all research objects, including data that are not (yet) used in research.

RESEARCH OBJECTS

In order to address the blurred boundaries between classical data and other elements of the scholarly system, Bechhofer and co-workers defined the concept of research objects [Bechhofer et al., 2010] as semantically rich aggregations of resources that can possess some scientific intent or support some research objective. I propose to use this very useful concept throughout this book in its broadest sense; research objects are defined in the context of this book as: any object, data element, or information, tools and services of many different kinds and levels of aggregation, and in fact any digital object that was the product of or is used in research. Even if we use the generic term data, in most cases this term will actually cover all research objects. According to Bechhofer et al., all research objects should be semantically rich and annotated, but a research object might not meet these ideal criteria and yet still be a research object. In addition, there is always confusion about data and software, algorithms, and services. In data-driven science, where the software and algorithms that make data actionable by machines are inseparable from the data themselves in operational terms, it also makes sense to treat code in the broadest sense, also as a form of (actionable) data. We will do that throughout this book. In this book, the term research object thus refers to *a collection of one or more digital objects produced by, or used for research*. In fact, we treat each (digitally operated) research object as a form of data.

MACHINE-ACTIONABLE

The term machine-readable is probably more common than the term machine-actionable, which has been associated with the FAIR principles. The reason the term machine-actionable was chosen as opposed to the more passive term machine-readable is that in open and data-driven science, the ideal situation is that (virtual) machines can relatively independently find, access, interoperate, and reuse data for the purpose they are designed for. So, machines should be able to quite literally *take action* on data, rather than just being able to read them.

FAIR PRINCIPLES

A basic premise of this book is that at the very highest level of dealing with data, with the ultimate aim to discover new knowledge, it is important that all research objects should be **F**indable, **A**ccessible, **I**nteroperable, and **R**e-usable (i.e. FAIR) [Wilkinson et al., 2016]. As a basic principle, hardly anyone will disagree with this, but of course we need to make implementation choices to actually make data and other research objects FAIR, with the R being the final purpose and the F, the A and the I being preconditions. There is an entire section about FAIR principles later on and we will see some examples of early FAIR-compliant implementations, but the *basic principle* should stick in our minds throughout the process of discussing and implementing data stewardship. If we see the collection of all linked digital data in the world for a moment as the *Internet of Data*, or rather *the Internet of FAIR Data and Services*[12], we can reduce the FAIR principles in effect to very similar simplicity as the basic principles on which the current Internet is operating. In fact, a robust routing to and connecting of digital research objects is what is needed. Very simple protocols (similar, but not identical, to TCP/IP) and unique, unambiguous, machine resolvable and persistent identifiers for each concept (*Persistent Identifiers*, PID[13]) are at the basis of FAIR.

MODERN SCIENCE

What we call *modern science*[14] is open science, a continuous scientific analysis process, based on an exploding amount of data generated in all segments of global society and where no longer only ivory tower scholars create and analyse these data. Rather, the entire society becomes *data-based*, and citizens/citizen corporations, such as patient societies, as well as organised private corporations and governments, increas-

---

[12]see HLEG report doi:10.2777/940154

[13]http://www.pidconsortium.eu

[14](just to avoid the further use of coined terms with specific connotation such as: Science 2.0, eScience, data-driven science, data-intensive science and big-data science)

ingly participate in the modern scientific process. In order to allow this enormous asset of a billion minds to participate in the global discovery and recording process, we need to be critically concerned with the quality of data, and particularly, arguably even more important than quality in itself an ambiguous term, with its *provenance*[15]. I (and my machines) need to know who generated the data, how, at what time, and for what purpose, to be able to judge whether I want to trust and reuse those data in my discovery or decision-making process. So, in modern science provenance is key. Later on we will read more about computers, but in modern science, by the very fact that it deals with data volumes and levels of complexity way beyond human reading and processing capacity, machines have become our major research assistants. So metadata (data about data or services) as well as the data itself should be machine-readable, and where possible, actionable, with non-scalable human intervention only where it is currently unavoidable.

Machines have become our major research assistants, so we had better prepare data for them.

OPEN SCIENCE

Open science is an umbrella term for a technology and data-driven systemic change in how researchers work, collaborate, share ideas, disseminate and reuse results, by adopting the core values that knowledge should be optimally reusable, modifiable, and re-distributable. One obvious and fundamental requirement for open science is thus that all research data and the associated tools and services should be FAIR [Wilkinson et al., 2016] (i.e. FAIR). However, securing the technicalities needed for optimal reuse is necessary, but not sufficient. The entire method of scientific research is in a landslide transition, so data need to be reusable in rapidly changing analysis environments, and may have to be regularly refreshed or parsed to different formats, which may introduce errors and ambiguities. So, at the basis of good open science is good data stewardship.

DATA STEWARDSHIP

So, what do we call data stewardship, and how is it different from data management?

---

[15]https://en.wikipedia.org/wiki/Provenance#Data_provenance

Data stewardship is such a key concept in this book that we need to define it up front. As we already saw, the term *data* is being used in different contexts throughout society. But in the context of research and discovery, we defined data according to a set of very basic principles, and thus the definition of data stewardship includes the good and long-term care of all research objects.

Data stewardship is now defined as:

The process and attitude that makes one deal responsibly with one's own and other people's data throughout *and after* the initial scientific creation and discovery cycle.

1.4   THE LINES OF THINKING

A first line of thinking throughout the book is that we need respectful collaboration of specialists as:

Trying to be a domain expert as well as a data expert will make you mediocre in both.

Science is becoming more and more interdisciplinary, and data from different domains need to be combined to discern the patterns that will lead to the major transformational discoveries of the future. Computers and (virtual) machines are now our key research assistants, so first of all, we had better make our data understandable for them. Experts who know computers, the programs that make them run, and the data we feed to them and receive from them should be highly valued expert colleagues in any modern open science research team dealing with complex data.

A second line of thinking is that a good data steward publishes findable, interoperable, accessible, and, therefore, optimally reusable (i.e.) FAIR (meta)data with a supplementary article (Data(+)), instead of the outdated approach (Article(+)), so that human-readable outputs and machine-readable outputs both have their rightful place in modern scholarly communication.

A third line of thinking is that all research objects, regardless of whether they are classical data, workflows, algorithms, or narrative articles, should minimally have FAIR metadata. In many cases, the actual data themselves can also be machine-actionable, but this is not a binary criterion for being considered FAIR or not. Soon, FAIR metadata and where possible FAIR data will be a prerequisite of many funders and publishers, but the main incentive to publish data and other research objects according to FAIR guidelines is that data and executive research objects (virtual machines, workflows, algorithms) become more interoperable and can be reused much more effectively, so that everybody benefits and the creators get proper credit. Connecting your data to all other FAIR data and services in the world can very significantly raise their value for knowledge discovery. The kernel of FAIR (meta)data publishing is:

Never refer to any concept represented in your data with ambiguous symbols or values.

Ambiguous lingual symbols are an even bigger nightmare for computers than they are for humans.

The book also deals at a very basic level with software, as from a data stewardship perspective all digital objects are data, and some digital objects are executable by instructed machines. So, throughout this book, all research objects that are the result of the scientific process will be covered under the umbrella term *data*.

The fourth line of thinking is that modern, data-driven science can only reach its optimal potential as open science, and a key feature of open science is the use of other people's data and services (here referred to as OPEDAS). Therefore, your own data and other research outcomes will hopefully be OPEDAS soon enough for your colleagues around the world, and that means that your responsibility is *not over* when you have your conclusions and your paper. Data should be a first-class citizen and should be independently machine-actionable whenever possible, but obviously, to make them also understandable and reusable for people, narrative text will still be needed when data are published. The reasoning behind conclusions and the rhetoric supporting claims are only partly prone to machine-encoding, and human reasoning as well as narrative rhetoric is not likely to disappear from science any time soon. However, by putting narrative centre stage in scholarly outcome communication, and due to clumsy data stewardship in past decades, many data are lost forever, experiments are not reproducible, and science has recently suffered a lot of negative press. Nothing a scientist sends into the world should be without solid evidence, rooted in solid reusable data. This is obviously even more important when scientific results can be translated in the short run into societal innovation. For instance, in health care, lack of good data stewardship may cost lives, and thus for data stewardship: *professional, or not at all* is the mantra.

In case you became interested in what all these terms mean, you will find some answers and guidelines in this book. Don't think this subject does not concern you, as no research data are excluded from the requirement for professional data stewardship.

1.5   THE BASICS OF GOOD DATA STEWARDSHIP

The two central hurdles for effectively using OPEDAS in open science are, in fact, *obscureness* and *ambiguity*. Many OPEDAS remain obscure, even if they would be reusable by others in principle. They are going unnoticed because they are in non-findable and ill-supported repositories for software or data, they are not adorned with rich and machine-actionable metadata, etc. This problem is addressed by several initiatives around the world and will not be a major topic in this book. However, the first rule of good data stewardship is obviously that you put your reusable stuff in a good repository and arrange for support. But even more alarming: Even if OPEDAS are found and accessible, the interoperability aspects form a second major hurdle and thus the final goal, reusing them, is still not met. So the second rule of good data stewardship is:

Never refer to any concept represented in your data with ambiguous symbols or values.

Bear with me, because if you give up or divert at this very basic philosophical level, the rest of the book will make very little sense to you. Let's reflect a bit more on the lowest granularity of a *persistent identifier*. If we define a *concept* as any unambiguous unit of thought humans can imagine (see [Mons and Velterop, 2009]), meaningful data, at their very core, are a set of symbols or tokens (characters, numbers) that refer to a meaning or a value and thus are composed of a set of one or more symbols that refer to concepts and their attributes/values in human minds. Humans have communicated about concepts and their relationships for ages using symbols (in sign language, spoken language, and narrative, and more recently in machine-readable and -actionable formats). Here we will not address philosophical issues on what exactly constitutes a concept and what does not. Nor will we discuss the good and the bad that has come from people communicating in ambiguous language and now in ambiguous computer languages since the *Tower of Babylon*[16].

The main message here is that:

Using ambiguous symbols in science should put you on intellectual Death-Row.

OGDEN TRIANGLE: THE RELATIONSHIP BETWEEN THE UNIT OF THOUGHT, THE TOKENS REFERRING TO IT, AND THE OBJECT IN REALITY

Computers have even more difficulty correctly resolving terminological ambiguity than humans do. Therefore, in the digital age, and certainly in data-driven science, each concept referred to in data needs a unique PID (supported by a human-readable definition to connect to our mental model). Just think of a concept as any defined unit of thought you can come up with regardless of whether this is a hard-reality entity, or a virtual concept, such as that referred to by the symbol for a given disease or the symbols for the concepts of *love* or *trust*.

As soon as we want to communicate scientifically about anything we had better have the best possible, jointly agreed definition of what we are talking about. If we want to ensure that *computers (which are much more binary than we are)* understand what we mean, we had better make sure that they are routed to the correctly defined meaning for any symbol we feed them[17]. So a PID is one thing, but pointing a machine to the correct place where this concept is defined is yet another thing, and the current Internet protocols require a form of a *handle* (usually a prefix with a unique post-fix, like a URL or a URI)[18].

But, let's stay a bit longer at the helicopter level. We can argue that a concept is a unit of thought in its own right and exists in our minds virtually, independent from any digital or physical representation of it. However, the very moment we assign a PID to a concept, that PID *itself* is a digital object and refers to the actual concept. A PID could therefore be regarded *de facto* as the minimal meaningful

---

[16]Genesis (5000 ? BC) 11:1-9, unfortunately no DOI available

[17]Please keep in mind that this is not much different from routing machines to the correct IP address or URI.
https://www.w3.org/DesignIssues/Meaning.html

[18]See continued discussions here https://www.rd-alliance.org/group/data-fabric-ig.html

and resolvable version of a digital object. Mind you, the acronym PSA in digital format is a digital object as well, but it is not unambiguously resolvable to a meaning. In fact, it refers to well over 100 different concepts in the literature. So, the concept *prostate specific antigen* should have a different PID from the concept *public service announcement*.

The first semantically meaningful aggregation level of digital objects is a set of PIDs representing a broader set of related conceptual elements, each with potential attributes. The recurrent minimal unit of one digital object can be aggregated to increasingly complex levels, and each aggregation is again a digital object. In this context, the smallest conceivable unit of assertional data is a single assertion, in science usually representing an association between multiple concepts, having the structure [subject] [predicate] [object] with its intrinsic and user-defined attributes[19].

By combining more and more complex and aggregated collections of digital objects into higher-level digital objects (assertions, graphs, datasets, databases, narratives, workflows, etc.) with a clear structure, a scalable system can be built that carries intrinsic interoperability. Ever larger aggregations of linked digital objects can create very complex digital objects up to complicated databases, and even the entire Internet could be seen a gigantic digital object. It is in this context that this book will generally refer to the largest conceivable interoperable data and services collection in the world as the Internet of Data and Services.

At this point, let's also check that according to the amateur-philosophical principles outlined above, any digital object is a form of data, and we could indeed view algorithms, software code, and any other research object also as data. So when we talk about the catchphrase Internet of (FAIR) Data and Services we in fact mean a global, interoperable web of data and related services.

All digital objects are data, some digital objects are executable by instructed machines.

Let's also be up-front about the limitations of science: It only tells us something about the *currently measurable* part of *reality*, whatever that is. Most of reality likely still escapes our observation and

---

[19]In RDF graphs, such as nanopublications, this relates to the assertion graph, the provenance and the publication graphs, each being intrinsic and user-defined attributes to the core assertion see www.nanopub.org

our ability to measure.[20] We discover seemingly reproducible patterns, and these are sensed as stable and reproducible. Our interpretation is that these appear as *true until falsified* or until we find a better and more insightful explanation for the same observed phenomenon. Thus, data, information, and particularly knowledge is in continuous flux and needs to be re-evaluated at very regular intervals. Therefore, individual concepts referred to in data may undergo *conceptual drift*, terms may undergo *semantic drift*, insights might change, rendering older data or their interpretation-at-the-time questionable, and long-established paradigms may crumble altogether.

Dogmatism about current knowledge stops progress both in science and in society.

So, this book itself should also not be dogmatic. However, we can be pretty sure that a very important element of good data stewardship is to mark, annotate, and/or throw away what is beyond its usability date, and that sometimes does not stop at data, but may include entire theories that have dominated a discipline for decades. Also, with the recent debates and publications about reproducibility of scientific results, fraud, and biased interpretations [Ioannidis, 2005], where lousy data management is frequently one of the underlying problems, we need to emphasise the critical role that proper data stewardship has to play in the repair of the damage done to the reputation of science. *http: //www. thenewatlantis. com/publications/saving-science.*

Thus, with the repeated nuance that we capture all research objects under the broad definition of data, we need to emphasise here, and I will do this throughout the book *ad nauseam* if needed, that:

Your responsibility for data is not over when you have your conclusions and your article!

By the very fact that we continuously *reuse* each other's prior research objects, we have a collective responsibility as modern scientists to treat our data as our core asset. That means that they are not disposable after they have served the purpose of writing a paper or, at best, kept as long as reviewers may wish to check them for reproducibility or fraud issues. More often than not, our data (including

---

[20]Until recently, LIGO observations would fall in that category.

our *negative* results) will be usable and reusable by many other scientists for integrated analyses and in disciplines we never even dreamt of when we created them. So if data are the *new oil*, the *new gold*, etc., why are we putting them in amateur repositories (graveyards) that are not interconnected, not findable on the Internet, not accessible, not interoperable and, thus, not reusable, or citable?

Most likely, because (a) the infrastructure to do all that is not there (or we don't know it is), (b) we do not know how to do it, and(c) the archaic reward system in the scientific ivory towers does not incentivise us at all!

Even today, there are still scientists who have no problem openly stating that they feel that they have all rights to keep data they created with public support to themselves for a variety of reasons [Sharing, 2016] and some even do not feel ashamed of calling people that reuse other people's data *research parasites* [Longo and Drazen, 2016]. The same people do not seem to have ethical concerns with the fact that they accepted taxpayer money in the first place in order to be able to generate their data. As we will see later, the FAIR principles clearly acknowledge that not all data can be open, but as a matter of principle, even if data are collected from citizens and they are to be considered privacy sensitive, the donor of the data should have the final say in how these data are used. The *collector* is at best a respectful custodian, and should also deal responsibly with the data from an ethical perspective, which is not by default keeping them only for themselves behind firewalls with the excuse of protecting their research subjects. New mechanisms to open up individual data at the request of the donors are pioneered in several settings, both in the private [21] and the public sector[22]. So, acting responsibly with data you generated with public funds, and recognising that you do not own the data privately, is a basic attitude for any data steward.

A researcher is the custodian of data, not the owner.

The starting point for data stewardship in this book is the perspective that knowledge generated with public funding is by *default* a public good, can improve society and save lives, and should therefore be reused for multiple purposes as efficiently as possible. Reasons for keeping data temporarily or perpetually in restricted access may exist,

---

[21]www.23andme.com/en-int/about/privacy/
[22]www.personalhealthtrain.nl

but need to be clearly spelled out in a data stewardship plan before a funder should decide to fund research from public funds yielding non-public data. So, for open science, open research objects are the norm and it is the responsibility of each and every researcher and funder to ensure not only publication, but also optimal reusability of all relevant data generated in the research process. It is not a *sinecure*, and in fact quite costly to make one's own research objects reusable for others, but it is part and parcel of good research practice, and the costs for it should be eligible to be included as part and parcel of research funding. Also, we foresee that many specialised public and private services will develop in new business models to support the Internet of FAIR Data and Services.

Thus, all research objects should be treated as first-class citizens in the science ecosystem, and the significant costs of open data should be seen as part and parcel of research funding and infrastructure.

Responsible data stewardship of each research object is part of good research practice.

This is where the FAIR guiding principles come in again: The final aim of data stewardship is not just to *preserve* the data, but to make them reusable, and more importantly, actually reused and acknowledged by others. To achieve this, in all data-related choices you make, you will need to consider whether these choices actually make data findable, accessible, interoperable, and, therefore, reusable. The FAIR guiding principles have been elaborated on in [Wilkinson et al., 2016] and we will address them in more detail later, but it should be clear right now that the FAIR guiding principles are not standards or protocols themselves. Rather, when you make implementation choices in the future, always consider whether your choices will render data FAIR.

Data stewardship (remember: in practice, research object stewardship) can thus also be specified as the good care of data throughout its full life cycle, from the planning for its creation, all the way through to its prolonged reuse. It is therefore clearly broader than data management within the boundaries of a particular research project. Data stewardship touches every aspect of data-intensive science and in fact is at the core of good research practice. Importantly, proper data stewardship is not to be seen as just a service to other, future users. First of all, the benefits will be reciprocal, as compliance with good research practices throughout the community will benefit all researchers that generate data, as they cross-talk seamlessly to other people's data (OPEDAS). Secondly, very few data that stand alone will be generated in the future, so the use of automated workflows for data analysis will be greatly enhanced when data and the associated executable code and workflows are *FAIR-borne*. That way, the contextual data that place newly generated data in their proper niche in the knowledge space do not have to be re-connected or re-asserted. This also means that once you publish your data according to FAIR and proper data citation principles[23], they will gain enormously in potential value.

As a scene-setting remark: Data stewardship is not necessarily more complex for the mythical big data[24] than it is for any other form of data. In fact, the complexity of data, for instance, the presence of many different (or rare) concepts to refer to, is more important than their sheer size. A very simple yes/no measurement over many different topical locations and dealing with only one measured object or phenomenon can generate petabytes of data that are in fact very easy to steward over time, while a relatively small but highly complex dataset that will be used across many different disciplines having update issues and semantic drift associated with it, can be very difficult to steward and reuse in a reproducible and citable way. So, let no one divert you with hype-arguments about *big data*. Valuable data needs proper stewardship, full stop.

Data stewardship starts with even the smallest data object.

So, what is a minimal data object that can be used for scientific analysis and discovery? And how do we steward it? Figure 1.17 depicts a highly simplified model of a meaningful data element. Please note again, the data elements box may actually represent executable code as well.

THE BUILDING BLOCKS OF DATA

As argued before, data are collections of digital objects, and even a single digital object such as a PID could be considered data. In research we usually deal with more complex data objects in which each

---

[23]https://www.force11.org/group/joint-declaration-data-citation-principles-final

[24]https://www.goodreads.com/book/show/18211094-big-data-at-work

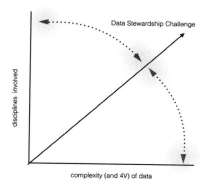

Figure 1.16 The increasing complexity of data, not necessarily its size, and the interdisciplinary character of datasets to be combined in a study, determine the nature and the magnitude of the data stewardship challenges in data-driven science.

dataset/slice is composed of *data elements*. For simplicity's sake, here we separate data elements into three main categories (although all of the elements are again composed of digital objects and could thus be considered data in their own right).

The minimal container that represents a bit of meaningful data contains the actual data elements, but in addition, data about those data elements and a PID for the container as a whole. Each meaningful bit of data is referred to here as a data element. Both in the metadata box and the actual data box, there could be assertions and associations. Metadata are in fact *assertions about the actual data*. The data container can be as small as a single assertion with its provenance and publication details (i.e., a nanopublication) with, for instance, a measured value (e.g., [subject] [measures] [1cm]) or a collection of data objects as big as a whole database (for instance, the 1,186,849 nanopublications representing protein locations in the Human Protein Atlas[25]). It can, however, also contain and represent a single image, a video, an article etc.

Obviously 1 and 0 type data do not *mean* anything if they are not placed in a context (*what measures 1 cm in this case?*). So data needs

---

[25]http://www.proteinatlas.org

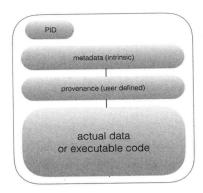

Figure 1.17 The simple building block of most data. The data element (as small as a single assertion, or as large as an entire database) needs to be described with rich and machine-readable metadata. The entire data container should have a persistent identifier to refer to. The metadata can be intrinsic (asserting facts about the actual data as correctly and richly as possible) or user-defined, which can be any annotation asserting more subjective things about the data, such as errors spotted, ways in which the data have been and can be used, where copies of the data can be found, etc. Typically, user-defined metadata will grow over time, especially when data or other research objects (there might be code in the data element) are intensively reused.

context and provenance. Context can be *intrinsic* to the data (person y went to shop x at time z), but in most cases, metadata will be needed to make sense of the data.

From a purely technical perspective, metadata are also data (data *about* data) and the difference is functional, not technical. So, data elements need metadata to place them in context and make them meaningful (in fact, now constituting *information*). These metadata can again be intrinsic, meaning that they are just factual assertions about the underlying data. Examples could be: file format, location, time-date stamp, method, and equipment used, etc. *User-defined* metadata can be much wider and more diverse; in fact, this form of metadata can contain anything anyone says about the actual data elements, at creation time or at reuse time. These can be assertions about *what is in this data container*, and, for instance, comments on how the original experiment yielding the data was conducted, and why the data can be (not) used for a particular purpose or with a particular workflow. It will occur to you immediately that the distinction between these two types of metadata will often be a gliding scale and thus debatable. Therefore, they are depicted as two shades of metadata, as there will be individual variations in decisions on whether, for instance, the *time that this data came into existence is intrinsic or user-defined*. One could argue philosophically that most data (series) are generated in a time *interval* rather than at a particular point in time. In any case, the actual data elements, let's say the elements on which one would perform deep analytics, need metadata. These will frequently be used (also by machines) for locating, accessing, and selecting relevant data, etc. In the early FAIR-compliant applications the metadata container is called a *FAIR accessor* [Rodríguez-Iglesias et al., 2016]. It allows FAIR-compliant search engines to discover data, and retrieve information about their character, licensing, accessibility, and reusability for particular purposes. The FAIR accessor can be updated over time, as will, for instance, be happening in the BD2K project CEDAR (Centre for Extended Data Annotation and Retrieval) [Musen et al., 2015]. A key long-term element of CEDAR is that whatever is asserted about an existing dataset (or workflow for that matter) over its life cycle time is considered an annotation and is thus part of the growing metadata about the research object. These assertions could also potentially cover statements such as *do not use this data in combination with this particular workflow*, or making explicit errors in the data.

WHAT EXACTLY SHOULD WE BE A STEWARD OF?

Data objects that play a role in research are in fact all research objects. Here is the place to briefly take you somewhat deeper into the concept of a research object. As stated before, a *research object* is defined here a bit more broadly than in the original paper by Bechhofer et al. [Bechhofer et al., 2010] as any artefact produced or used in the scientific process[26]. Data objects as defined above can clearly be research objects, but also executable code, virtual machines, images, slides, samples, and classical narrative articles are research objects. A researcher may rely on existing (other people's) data that were or were

---

[26]http://www.researchobject.org/

not purposely created by an experiment, such as social media streams, newspaper data, outputs from wearables, or generic climate measurements. Researchers actually also increasingly rely on other people's prior algorithms, software and protocols, up to Web services, APIs, as well as ready-to-run virtual machines. So, all these research objects should not only be carefully stewarded, but also in many cases their reuse should be supported. In the current situation there is very little incentive for science professionals and active researchers to actually support the reuse of their data or their services by others, as there is no established culture to properly cite, recognise, and reward such activities. This is a typical role for data stewards and not necessarily for active experimentalists. The data and software quality (or lack thereof) is a challenge re-users have to face on a case-by-case basis. The quality and sustainability of published code, workflows, etc. in the academic setting is a serious concern, also to be addressed later. However, if you generate *de novo* data or code purposely for research in a given experiment, it is the worst malpractice thinkable in modern science (apart from conscious fraud, maybe) to generate and disseminate these research objects without a good stewardship plan. After all, science is not just a creative art, it is also a profession with profound influence on society.

Nothing a scientist sends into the world should be without solid evidence, rooted in solid reusabledata.

WHY SHOULD WE STEWARD OUR RESULTS IN THE FIRST PLACE?

Conclusions may be wrong, and most conclusions will sooner or later appear to be wrong or only partially right, but the evidence (data + analysis = prior art) on which they were built should be traceable, and where possible, literally reproducible. An observation on the hype-term reproducibility should be inserted here. If results are not reproducible, the conclusions are not necessarily *wrong*. There may be many unknown reasons why results from earlier (particularly older) experiments cannot be exactly reproduced years later [Ioannidis, 2005]. However, utmost care should be taken to support reproducibility of experiments to the greatest extent possible. Still, non-(exactly)reproducible data-based analytics need to be carefully evaluated for their likelihood to have led to the correct conclusions. Reproducibility issues in open science go way beyond the reviewers of an article being able to check this aspect. Workflows running on the same data should repeatedly give the same results. This is currently not the case in many instances, which can be due to updated components in a workflow (for instance, the thesaurus supporting a text-mining tool) or unreported changes, corrections, or updates in the data itself. Scientists are notoriously sloppy in exploiting a proper version management system around their experimental tools and data. So, data stewards need to be acutely aware of this common source of irreproducibility alarms and do everything to prevent their experimentalist colleagues (including data scientists producing new code) from causing confusion with undocumented updates on code, APIs supporting components, and data. Once the OPEDAS assets become routine in open science practice, unexpected changes in components of a repeated workflow will become a major problem, unless proper data stewardship prevents this. Currently, common practice in science is way below the desired standards for open science. Reproducibility is severely hampered by data stewardship malpractices, which is for instance the case in many studies, even in clinical medicine [Freedman et al., 2015]. This is partly due to the fact that in 2015, as many as 88% of biomedical research datasets were not demonstrably deposited in a well-known, public repository [Read et al., 2015] and will effectively simply get lost for the commons. Let that sink in, and recall the shocking findings reported in [Read et al., 2015], claiming that in the United States alone, an astonishing $28,000,000,000 per year in preclinical research is not reproducible.

Lack of good data stewardship may cost lives.

IS DATA STEWARDSHIP AN ACADEMIC ISSUE ONLY?

Before we really start, let me address one more issue, as it causes frequent debate and misconceptions: the relative role of *public* and *private* parties in data stewardship. There is a strong tendency in academia to take care of our data ourselves. This may be a result of our traumatic experiences with the current narrative, largely monopolised publishing model. It may also be routed in a deep but frequently ill-rationalised general *anti.com* mentality in academia. Meanwhile, in day-to-day reality, we trust our personal data and our personal safety in the hands of private parties all the time, at least every time we fly (as I did most of the time when I produced this text). We also know very well that reliability, supportability, and, consequently, sustainability are usually not terms we associate strongly with funding-cycle-dependent public-sector projects, and not even with specialised public-sector institutions. According to Barnes et al. [Barnes et al., 2009], high-quality, open, and accessible data are the foundation of pre-competitive research, and strong public-private partnerships have considerable potential to enhance public data resources, which would benefit everyone engaged in, for instance, drug discovery. This is a nice way to say that the current system, with in house bioinformatics solutions relying on non-interoperable and unstable external public resources, is broken. So, how do we find the right balance between highly challenging data analysis and stewardship issues that need to be addressed first-off by academic groups, and commodity software, hardware, and databases to be used commercially off-the-shelf for more established data stewardship goals? For the latter category, re-inventing the wheel is probably one of the most ineffective and disturbing practices[27], and results in too many standards and too many solutions to choose from, in turn leading to fragmentation and non-FAIR-ness.

PROFESSORWARE

Research objects (especially research data and code) are created almost by definition initially by active researchers who need them, regardless of whether they are in industry or in academia[28]. The most advanced data, formats, and code are usually pioneered by academics for scholarly purposes, but increasingly, researchers and developers in private and industrial environments also create data and code for internal purposes to begin with. It is very natural, and actually a practice we need to consciously preserve, that considerations about performance-optimisation, scalability, and professional support (i.e. service level agreements) do not fully come into focus at this early stage. There is an emerging trend to teach academics the most basic approaches for software carpentry and data carpentry, but we should not try to make scientific prototyping engineers focused on scalable and sustainable software that is industrial grade, as it may kill their creativity and out-of-the-box thinking. Let's call the first phase of research objects that address a novel intellectual challenge *professorware*. This is

---

[27]A must-view: http://www.slideshare.net/dullhunk/the-seven-deadly-sins-of-bioinformatics

[28]Reiteration: data not generated with research in mind are the exception and pose a set of specific challenges.

in no way a derogatory term. We need professorware in order to make more than incremental progress.

My favourite metaphor to explain the various phases in the process from attacking the intellectual challenge to a professional, maintainable, and thus sustainable software product is that we first need to *hack a path of a few meters into the jungle*. We show that there could be a way, and we show the approximate direction. At this stage, professorware is the way to go. Once we know that the direction of the solution is intellectually feasible, and we have a working prototype, we need to consider the next steps. The next phase is calling in the bulldozers. Before anyone will finance sending in the bulldozers and building an asphalt road, investors will want to have answers about market (how many travellers will use the road, and for what purpose?), scalability, supportability, and, thus, long-term sustainability. However, if the hackers of the first few meters would be bothered by that, they may either never have started or they may have made the wrong decisions. Obviously, good software development practice is also already important in the professorware stage, and if academic code is well structured and documented, it becomes a lot easier to turn it into a sustainable product once it shows initial value.

Another reason to separate professorware (also organisationally) from industrial-grade solutions is the tendency to destroy a (spin-off) company by making a creative scientist CEO. Once the scientist has made a black-and-white television, (s)he has probably already conceived of the colour television. Instead of first selling everyone a black-and-white television, and then selling them a colour version once they got addicted, the scientist would tell potential customers to wait for the colour television. The two major factors that would destroy the company are, first of all, that it would miss the opportunity to recover some development costs on the black-and-white television, but probably, more importantly, that it is law of the Medes and Persians that actually getting the colour television to work appears to take much longer (usually a factor Pi) than the scientist or science engineer anticipated once the intellectual nut is cracked.

So, for a good data steward, I recommend the following position (also towards data scientists and experimental scientists in the group):

- Professorware is a crucial step in the process of disruptive change and in many cases perfectly serves the exploratory aims of the researcher.

- Considerations of scalability, performance, addressable market, and consequently, sustainability, should not get in the way at this early stage.

- These questions should, however, be answered satisfactorily before professorware is turned into a sustainable system or component, and published or presented as testable and reusable by others. This step frequently means going back to the drawing board with professional architects and engineers while the scientist remains on board as a content and functionality consultant.

- As argued before: in e-Science, data and processing code cannot be meaningfully separated, so these lessons hold as much for data-preserving code as they do for data-processing code.

So where are the big pitfalls in all this? The big pitfalls are slightly different for results of scientific experiments, for data that emerge without a specific research purpose at their basis (for example, Twitter streams) and for processing code. To start with the latter: It is obvious that software carpentry and good software development rules in general are useful, including in academic software development settings, and will make refactoring easier in later stages. However, here the freedom of creative thinking is extremely important, and enforcing to much good industrial practice on creative hackers may actually hamper the development of disruptive professorware. But a major valley-of-death phase in the process towards a sustainable system is the failure to recognise or even the conscious denial on the part of the scientific programmers that they make professorware that in many cases will fail on minimally one aspect that would make it sustainable. There are obviously exceptions where software that was made by academics for scientists has made it into long-term use by a growing community of users and contributors/improvers without a professional SLA service provider, or continue to function in a support environment such as CERN, EBI, or NCBI. I would argue, however, that such examples (apart from being the exception that confirms the rule) are relatively simple and straightforward single-purpose algorithms with very simple interfaces that are being used by highly specialised people who have no alternative and consequently accept academic best practice performance. Again, this is a very important subset of processing code and it will be with us to stay. However, things go astray when scientists and even academic engineers claim that what they build is sustainable, industrial grade, 24/7 reliable, and ultimately scalable. Again there may be exceptions, but the rule is that professorware is none of these things. No problem, unless scientists effectively get in the way of allowing professional architects and engineers to address these issues properly and redesign what the scientists initially built. Take as one example the many workflows collected in Taverna, as such an open source development infrastructure hosted by apache.org[29]. As shown by Mayer and Rauber [Mayer and Rauber, 2015], the majority of the workflows exposed in Taverna are not re-executable, and often the cause is to be tracked back to rather trivial shortcomings, such as the lack of example values needed as workflow input parameters. Also, missing libraries for Java programs are frequent causes for failures in workflow execution. This is obviously not a shortcoming of the Taverna framework itself, but if workflows do not have a frequent use, are not regularly re-run as a test on reference datasets, and are supported only by a PhD student who naturally moves on, and have no service level agreement (SLA) attached, while the code is probably also sub-optimal in terms of performance, scalability, and supportability aspects, it stays professorware and will likely soon die in beauty. The danger is that some of these workflows are still built into sequential workflows and Web services, and cause major problems when they are down or outdated. The conclusions of the Workflow4Ever project speak for themselves in this respect. For relevant publications on workflow sustainability aspects, see: http://www.wf4ever-project.org/publications.

DATA AS A FIRST-CLASS CITIZEN

As emphasised before, as a rule, data are now too large and/or complex to be processed by reading. Therefore, software and algorithms to process data cannot be treated as independent from the data they process. However, can we tolerate the same compromises and margins for data as we just discussed for professorware? While professorware as initial data processing code is frequently crucial, experimental and other research data [30] as the central core substrate of modern science cannot be any less than as optimal as possible. Although one might argue that there is a time window for suboptimal data in terms of

---

[29]http://www.taverna.org.uk

[30]Please note that we distinguish experimental data from rescued data, the latter being not purposely created for research purposes. Here we talk about the direct outcome of your scientific experiments.

the lack of externally understandable documentation when data are only used for internal purposes, data that are intrinsically suboptimal should be banned from science as much as possible. In many cases, data that were originally generated for local use only may appear useful in further experiments and analytics carried out by others. Therefore, data stewardship starts at the phase when an experiment is planned. For instance, failure to capture rich-enough metadata to make data reusableby others may render it very difficult or impossible to offer the data for reuse later. Even if experimental software is used, because there simply is not yet any stable software available, the fact that such unstable services and algorithms have been used in the data processing and interpretation process must be well documented in the metadata (provenance) description. So, at least the source for non-reproducibility may be found in the workflow rather than the data.

FAIR DATA PUBLISHING

Achieving the transition from the current closed and siloed approaches to research towards more open and networked scholarship needs more than just changes in current practice. It needs a support infrastructure of data platforms, analytics, computational capacity, virtual machines, and workflow systems. Data formatting and publishing approaches (regardless of the schema and format chosen) should follow the FAIR principles. The FAIR principles do not demand any particular format or standard. They simply ask from you that you take optimal care of the four elements of the FAIR principles. Please recognise once more before studying the principles below that they may only pertain to the metadata or the annotations of otherwise non-machine-recognisable or -actionable objects (see Figure 1.18). In that case a unique and persistent identifier should be assigned to the physical object (a sample-tube, a piece of text, a picture, a person) and the metadata should be persistently linked to the physical object, so that both the metadata and, through those, the digital object they refer to, become findable and accessible for reuse, even if there may be several steps in between, before the actual resources or data become interoperable, linkable, or integrated for the study for which they are needed. This may even include reanalysis, further measurements (such as looking for new metabolites in an old sample that were not measured before), but the fact that the object that can be a relevant source of data is available under well-defined conditions for reuse in research should be adequately described in the metadata and annotations associated with the object.

To be Findable:

F1. (meta)data are assigned a globally unique and persistent identifier
F2. data are described with rich metadata (defined by R1 below)
F3. metadata clearly and explicitly include the identifier of the data it describes
F4. (meta)data are registered or indexed in a searchable resource

To be Accessible:

A1. (meta)data are retrievable by their identifier using a standardised communications protocol
A1.1 the protocol is open, free, and universally implementable
A1.2 the protocol allows for an authentication and authorization procedure, where necessary
A2. metadata are accessible, even when the data are no longer available

To be Interoperable:

I1. (meta)data use a formal, accessible, shared, and broadly applicable language for knowledge representation
I2. (meta)data use vocabularies that follow FAIR principles
I3. (meta)data include qualified references to other (meta)data

To be Reusable:

R1. meta(data) are richly described with a plurality of accurate and relevant attributes
R1.1. (meta)data are released with a clear and accessible data usage licence
R1.2. (meta)data are associated with detailed provenance
R1.3. (meta)data meet domain-relevant community standards

The wide embrace of the FAIR principles by governments, governing bodies, and funding bodies, has led to a growing number of data resources attempting to demonstrate their FAIRness. Examples can be found in [Mons et al., 2017] from which the following text is adapted. In some cases, however, the original meanings of findable, accessible, interoperable, and reusable may be stretched, sometimes even as a means of avoiding change/improvement. In other cases, the proposed means of implementing the principles that would lead to an Internet of FAIR Data and Services raises concerns and confusion. Therefore, in the context of this book, it seems valuable to clarify the principles in more detail for those who actually wish to implement FAIR-compliant data and or services. Data publishing infrastructure needs are being addressed intensively at the EC level, especially in the context of the 2016 Dutch EC Presidency and the European Open Science Cloud (EOSC[31]), and in the United States through the NIH Data Commons[32]. Comparable efforts are under development in Australia, Africa, Latin America, and Asia. Common to all these is the idea of building platforms that support the sharing of resources. Provision of all such resources and platforms will necessarily involve a mix of players, including commercial and public resources, and thus the FAIR principles are relevant to all these global efforts. Ensuring that all provisioned resources are findable, accessible, interoperable, and reusable, as well as ensuring that the qualities of a service (i.e., what it does, and how) as well as the quality of a service, are appropriate for the researchers' needs, requires widely shared and adopted standards and principles.

WHAT FAIR IS NOT

FAIR is not a standard: The FAIR guiding principles are often incorrectly referred to as a standard, even though the original publication explicitly states they are not [Wilkinson et al., 2016]. The guiding principles allow many different approaches to render open science and services findable, accessible, interoperable, to serve the ultimate goal: the reuse of valuable research objects. A standard would specify the means of implementation, which is overly prescriptive and hinders uptake. FAIR simply describes the qualities or behaviours that would be required of open science resources to achieve, possibly incrementally, their optimal discovery and scholarly reuse.

FAIR is not equal to RDF, Linked Data, or semantic Web: The reference article in *Scientific Data* [Wilkinson et al., 2016] emphasises the machine-actionability of open science and/or metadata. This implies (in fact, requires) that resources wishing to fulfil the FAIR guidelines must utilise a widely accepted machine-readable framework for open science and knowledge representation and exchange. While there are only a handful of standards and frameworks that could, today, fulfil this requirement, other potentially more powerful approaches may appear in the future. As such, FAIR explicitly does not equate with the use of for instance the well known Resource Description Framework (RDF) or any other semantic Web framework or technology. However,

---

[31]http://ec.europa.eu/research/openscience/index.cfm?pg=open-science-cloud
[32]https://datascience.nih.gov/commons

RDF, together with formal ontologies, are currently a popular solution to the knowledge-sharing problem that also fulfil the requirements of FAIR. As such, RDF + widely adopted ontologies figure prominently in many of the early FAIR examples[33].

That said, we should recognise that RDF has clear limitations today when it comes to high-performance analytics over large open science sources. It is therefore very likely that applications in the Internet of FAIR Data and Services will use a variety of open science formats that allow specific and scalable manipulations of open science for pattern recognition and knowledge discovery. It is thus important for professional open science stewards to keep discussions on FAIR-compliant formats precise and not suggest that any one format is a panacea. RDF + proper ontologies are very effective as a static interoperability format, but other formats may also be used in FAIR context, and high-performance analytics applications may use yet very different formats.

FAIR is not assuming that (just) humans can find, access, reformat, and finally reuse data: In a very instructive blog about the FAIR principles from Wageningen University, the following was stated eloquently: *"The recognition that computers must be capable of accessing a data publication autonomously, unaided by their human operators, is core to the FAIR principles. Computers are now an inseparable companion in every research endeavour"*[34]. As argued, time wasted on (repeated) munging may vary between 50 and 70% in data-intensive research. In the hypothetical situation that researchers and their machine-assistants would only have to deal with FAIR data and services, this time-waste would be reduced to a minimal fraction of what it is today. To serve this potentially enormous cost reduction, FAIR-compliant (meta-) data and services should be actionable by machines without human supervision wherever possible.

FAIR is not equal to open: The 'A' in FAIR stands for *accessible under well-defined conditions*, while reusability conditions are covered in the requirement to have a clear, machine-readable licence as per the R of FAIR. There are legitimate reasons to shield data and services from public access. These include personal privacy, national security, and competitiveness. The FAIR principles, although inspired by open science, do explicitly and deliberately not address moral and ethical

---

[33]http://www.dtls.nl/FAIR-data/FAIR-data/FAIR-data-knowledge-expertise/
[34]https://www.wur.nl/en/newsarticle/FAIR-guiding-principles-published-in-journal-of-the-Nature-Publishing-Group-family-.htm

issues pertaining to openness of the data. In the envisioned Internet of FAIR Data and Services, the degree to which any piece of data is available, or even advertised as being available (via its metadata, is entirely at the discretion of the data owner. FAIR only speaks to the need to describe a mechanised or manual process for accessing discovered data; a requirement to openly and richly describe the context within which those data were generated, to enable evaluation of its utility; to explicitly define the conditions under which they may be reused (if any); and to provide clear instructions on how they should be cited when reused[35]. None of these principles necessitate data being open or free. They do, however, require clarity and transparency around the conditions governing access and reuse.

FAIR is not a life sciences hobby: The first definition of the FAIR principles came from a group that was mainly perceived as coming from a life sciences background, and ELIXIR[36] was one of the first to adopt them. However, the principles may be equally applied to any data, or any service, in any discipline. The problems that hinder data reuse in the life sciences—ambiguity of symbols, too many persistent identifiers for the same concept, semantic drift, and linguistic barriers, the description of analytical methodologies, tools, and their capabilities, and the need for adequate and accurate citation—are all, in various shades of severity, also problematic in other domains, such as the humanities or law.

IS FAIR FAIR?

The actual meaning of the word *fair* in daily life is in some ways also confusing, as people have different perceptions and connotations associated with it. One major criticism (relating to the machine-actionability aspect of the principles) is the connotation that non-machine-readable data would be considered in some way *unfair*. It should be explicitly stated that FAIR is a continuum. There is no such thing as *unFAIR* associated with the FAIR principles, except maybe data that are not even findable. Obviously, not all data or other research objects can be machine-actionable. Moreover, there are circumstances where making data machine-actionable would *reduce* their utility (e.g. due to the lack of tools capable of consuming the machine-actionable format). As

---

[35]See the FORCE11 Joint Declaration of Data Citation Principles: https://www.force11.org/group/joint-declaration-data-citation-principles-final

[36]https://www.elixir-europe.org

such, we emphasise that as long as such data are clearly associated with FAIR metadata, we would consider them fully participatory in the FAIR ecosystem. However, a very positive connotation aspect is that the FAIR acronym carries the ring of general *fairness*. On the one hand the 'A' allows FAIR shielding of data that cannot be open for good reasons of all kinds, so industry and medical researchers are assured of privacy protection. On the other hand, the basic principle that FAIRness is maximised when data are open, maximising accessibility implies maximising openness. This includes addressing, to the greatest extent reasonable, the machine-actionability aspect of FAIR. So, the FAIR connotation should neither be underestimated.

PARTLY FAIR IS FAIR ENOUGH

See for reference to categories of FAIRness Figure 1.18. A minimal step towards FAIRness is to adorn the dataset, as a whole, with a PID that is not only intrinsically persistent, but also persistently linked to the dataset (research object) it identifies (B). We here distinguish intrinsic metadata and user-defined metadata. The former category (with the boundaries sometimes blurred) are the metadata that should be constructed at capture. In other words, they describe the metadata that is often automatically added to the data by the machine or workflow that generated the data (e.g., DICOM data for biomedical images, file format, time date stamps, and other features that are intrinsic to the data). Such data can be anticipated by the creator to be useful to find, access, interoperate, and, thus, reuse the research object.

As it is very cumbersome to peer review the quality of large datasets at the time when they are first published, the ongoing annotation of data sources during the period of their existence and reuse is a crucial process in open science, so we argue that both intrinsic and user-defined provenance (e.g., contextual) metadata should be added, and made FAIR whenever possible (C). Not all data lend themselves to be machine-actionable without human intervention (some raw data, but also images, for example). However, many data that have a relational and an assertional character can be captured perfectly correctly in a machine-processable syntax and semantic. Still, even if data are technically FAIR, it may be necessary to restrict their access for reasons discussed elsewhere,(D). However, the default for maximal FAIRness should be that the data themselves are made available under well-defined conditions for reuse by others (E). I argue here that even the step from A to B would already have a profound effect on the reuse

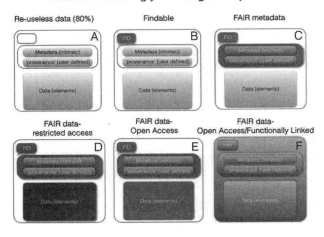

Figure 1.18 Varying degrees of FAIRness. As elements become more shaded (coloured in the e-book), they become FAIR. For example, adding a persistent identifier (PID) increases the FAIRness of that component. Coloured elements in green are FAIR and open, coloured elements in red are FAIR and closed. In the final panel, the mechanism for expressing the relationship between the ID, the metadata, and the data is also FAIR (i.e., it follows a widely accepted and machine-readable standard, such as DCAT or nanoPublications) and interlinked with other related FAIR data or analytical tools on the Internet of FAIR Data and Services. (With permission from Mons et al., Information Services & Use, vol. 37, no. 1, pp. 49, 2017.)

of research objects, because at least they can be found, and relocated, by those who know the identifier, which constitutes a minimal degree of FAIRness. However, thereafter, the addition of rich, FAIR metadata is the first major step towards becoming maximally FAIR. When the data elements themselves can also be made FAIR and opened for reuse by anyone, we have reached a high degree of FAIRness, and when all of these are linked with other FAIR data, and accessible to FAIRdata-compliant services annotated with FAIR metadata themselves, we will have achieved the Internet of (FAIR) Data and Services (F). However, when data are not FAIR (at least at the C level) they simply cannot participate in this vision.

The FAIR principles have sparked a serious debate about better data stewardship in data-driven science, but they have also triggered funders' requirements and thus implementation discussions; some of these are very embryonic, while others have matured into working prototypes.

Obviously, the FAIR principles are not magic, nor are they presenting a panacea, but they guide the development of infrastructure and tooling to make all research objects optimally reusable for machines and people alike, which is a crucial step. It is very important that the community will continue to discuss, challenge, and refine their own implementation choices, within the behavioural guidelines established by the principles.

The transparent but controlled accessibility of data and services, as opposed to the ambiguous blanket-concept *open*, allows the participation of a broad range of sectors, public and private, as well as genuine equal partnership with stakeholders in all societies around the world.

NOW, ON TO THE REAL QUESTIONS

I hope that after reading this introductory section, you became convinced that studying the rest of this book is crucial for your further scientific career. Again, if you are or want to be an experimental scientist, do not try to use this book to become a half-baked data expert, but use it to learn how to respect, talk to, and collaborate with real data professionals. If you are or want to be a data expert or, beware, a data steward, please do not expect that you will be a *would-be data scientist* after reading this book. However, use it to learn how to respect, talk to, and collaborate throughout the data cycle with domain experts as well as with hard-core data and computer scientists. Take the time to read about the basics of what the domain experts have to deal with, but do NOT try to become a half-baked biologist, chemist, geologist, or social scientist. You will fail on both fronts to reach the A-status needed to get funds. Share your expertise and collaborate with deep and lasting respect. If you are an old, established leader in science, or science policy, still continue and at least skim over the content of the next sections, in order to convince yourself of the daunting complexity of good data stewardship, so that you run downstairs and give data stewards tenure.

No areas or levels in research are excluded from the responsibility of data stewardship.

CHAPTER 2

Data Cycle Step 1: Design of Experiment

Before you decide to embark on any new study, it is nowadays good practice to consider all options to keep the data-generation part of your study as limited as possible. It is not because we can generate massive amounts of data that we always need to do so. Creating data with public money brings with it the responsibility to treat those data well, and (if potentially useful) make them available for reuse by others. There is considerable effort and cost associated with making data FAIR, and generally speaking, recreating data that may exist somewhere else is a waste of public resources. So, given the research question you would like to address, the very first question in open science setting should always be:

2.1 IS THERE PRE-EXISTING DATA?

What's up?

For many decades if not centuries, virtually every experiment started with the collection or creation of observations, and, in fact, data. In social sciences and humanities, the tendency to reuse data that had been created earlier, in all kinds of surveys and increasingly, of course, from sources such social media, may be already somewhat more established. However, in many of the hard experimental sciences, the generation of new data specifically produced to answer a hypothetical question is still so commonplace that careful thinking about the actual need to generate new data may just not be on the radar screen. Obviously, data creation will need to continue, but increasingly we have to ask the question whether such new data are absolutely necessary to answer the question we want to answer. With more and more data becoming available in reusable format, there may well be existing data collections of other people's data and associated services (OPEDAS) that with or without some extra effort, can answer at least part of the question or at least may be crucial for the interpretation of your own data.

DO

- Search for datasets (OPEDAS) that may be reusable and can help you reduce the number of new datasets you may have to generate (and steward later on).

- Include annotated collections of data and curated databases in your search.

- Check the accessibility and license situation attached to the relevant datasets you found.

- Check their interoperability. They may be relevant but not interoperable with your analysis pipelines. In that case, you may have to extract, transform, and load (ETL) them, or decide that - although relevant - they are not reusable for your purpose.

- Ensure that using OPEDAS will not restrict in any way the use of your results later on, including copyright and freedom to operate on the request of IPR.

- Check how to cite and acknowledge OPEDAS.

- Consider how to actively involve OPEDAS owners in your research, in order to make optimal use of their data.

- Speak to colleagues who did similar experiments before, to find out about potential OPEDAS you may consider using.

DON'T

- Assume no OPEDAS exist without thoroughly checking and using all your possibilities.

- Start an experiment without properly checking with colleagues about the best approach and OPEDAS out there.

- Budget for data generation in your study without justifying to the funder why the generation of the data is necessary.

- Move into actual experimentation without consulting a data expert.

Resources: http://dmp.fairdata.solutions/resources/atq

2.2   WILL YOU USE PRE-EXISTING DATA (INCLUDING OPEDAS)?

What's up?

Even if OPEDAS appear to exist, it is not a given that you will automatically use them. There might be considerations as discussed under Section 1.1, but it may also be just too cumbersome to get the data, or get them in a format that is easy to use for your particular study. However, if the answer is yes, there are a number of basic rules you need to stick to (after checking the features listed in Section 1.1).

DO

- Consider whether you need 'all' the data - some sets are very large and expensive to download and host - or just a relevant subset.

- Determine whether you need to download the OPEDAS or whether you can use them where they are (for instance, sending a process virtual machine or a workflow to the data).

- Check the reuse (license) and citation/acknowledgement policy provided by the data owner.

- If none is provided, contact the data owner to check these issues.

- Make sure (with a second opinion) that using the data, even for small parts of your analysis, does not restrict you in publishing or using your results later on. If unclear, contact the data owner.

- Actively annotate the dataset used if there is any issue with it, and submit these new metadata to a public, trusted repository.

DON'T

- Ever use OPEDAS without properly citing them in your resulting publications.

- Use data without a proper license. Even if they seem entirely open.

- Download unnecessarily large (portions of) data and host them locally.

- Store data locally for longer than necessary (assuming the original repository is sustained).

- Change anything in the downloaded data or its metadata without proper documentation and annotation.

- Move into actual experimentation without consulting a data expert.

Resources: `http://dmp.fairdata.solutions/resources/ezi`

2.3   WILL YOU USE REFERENCE DATA?

Reference data are defined as OPEDAS that have a status as a 'reference' dataset and can be used to 'interpret' other data. Reference data can be available in many formats and could include curated resources like LITMED in the humanities, UniProt or PDB in the life sciences, but also resources like a '5 year Twitter trend', a 'gold standard,' for instance a human reference genome that you use to define how your data fits in the larger picture.

What's up?

Reference datasets become more and more important when relatively large new datasets are generated and need interpretation. But, also, for small data, such as, for instance, a clinical genetics sample with hundreds of variants to be checked against what is known about their phenotypic associations. In some cases (like in ELIXIR) these resources may be branded as 'core resources'. These typically are dynamically updated resources, so they cannot be expected to be identical at different times of download or consulting. Increasingly, pre-analysed or pre-linked data sources will also enter the realm of core resources and reference data. When these are not offered by authorised providers, with a clear versioning and release policy, it is important to double-check the provenance and the ability to trace them back to the original resources, as well as the transparency of the methods used.

DO

- Make sure you understand the full range of reference data sources available that have potential relevance for your study.

- Check the 'authority' of these databases and be sufficiently critical about their validity.

- Make sure the conditions of use suit your purpose (some of these, like HPA, formally forbid, for instance, commercial use).

- Ensure that at a later stage you will be able to reproduce the data reused in your study as exactly as possible, and at a very minimum, record the version of the reference data resource you used.

- If you have doubts about reproducibility issues, consider downloading the data you cite and archiving them properly.

- Look for existing workflows or methods to (semi-) automatically check your data against reference data, to avoid building workflows that already exist.

DON'T

- Restrict your search for reference data to the obvious discipline-specific resources.

- Assume that all reference data (including peer reviewed literature) is correct.

- Use non-authorised reference data unless authorised, public reference data is available.

- Use reference data without proper citation or under other people's accounts, as it is important for reference resources to track their user statistics properly.

- Refer to reference data sources that are updated without a version number.

- Cite data resources or services that are obviously updated, meaning that people trying to reproduce your results will be led to different versions.

Resources: `http://dmp.fairdata.solutions/resources/quc`

2.4 WHERE IS IT AVAILABLE?

What's up?

Datasets (including reference data) may be available at different locations (in replica) and with different service level agreements attached. Established 'core' resources usually have a policy and a sustainability plan to ensure that the data will be available and properly versioned for a longer period of time. Smaller datasets and databases may not have such crucial features secured, which poses a strong risk. Especially when you decide to use data resources outside of your immediate domain, this can be an easily overlooked issue. It is wise to use those resources that are most effectively accessible and sustainable. Data that are available in a repository that is not approved, do bring additional risks. If you use these data for your experiments and they are no longer available later on, because, for instance, the repository went off line or got closed off, you may get into trouble with respect to follow up experiments or reproducibility and review issues.

DO

- Make sure you use data 'online' only when there is optimal insurance that the data will remain available (under the same conditions) 'indefinitely-in-principle'.

- If you decide to use data from a non-authorised resource, make sure you download and keep the entire data file locally, with proper documentation on the provenance.

- Check the performance issues related to using data 'online' versus 'locally'.

- Check all steps in intended workflows to be used in your data analysis down the line, and whether they support the data formats and availability these workflows support.

- Make sure you have (access to) the capacity to develop ad hoc and custom workflows where existing ones may fail, given the intended analytics procedures.

DON'T

- Assume that all OPEDAS resources you use 'online' will automatically be there again (and in the same format) the next time you need them.

- Use OPEDAS from a non-DSA resource if the same data are also available in a DSA version.

- Use services that have no SLA or sustainability plan, as this will jeopardise the reproducibility of your research.

Resources: `http://dmp.fairdata.solutions/resources/ckt`

2.5 WHAT FORMAT?

What's up?

Several datasets, and, in particular, reference data resources may be available in different formats. For instance, core resources in the life sciences, such as UniProt and the Human Protein Atlas, increasingly offer their core data in machine-readable (FAIR) formats. You need to be fully aware of the data formats used, the limitations of the format, the possible license restrictions, and thus the way in which you may or may not be able to reuse these data. Please be aware that even for data resources that are not password protected in any way, and can be freely used at face value for academic purposes, there may still be restrictions on reuse.

DO

- Check the data formats available at the sites of the OPEDAS sources you have selected to use.

- Make sure to pick the best format suited to the analysis you intend to do in your study.

- If the desired format is not readily available, consider contacting the resource owner to discuss whether the data can be made available in that format.

DON'T

- Use sub-optimal data formats without checking the availability (maybe at other sites) of the correct format.

- Use data for a particular purpose without making sure the format and the potential restrictions will not jeopardise your analysis or your abilities to publish.

- Use data or resources that are only for academic use if you intend to use the results later for potential innovation of commercial purposes.

Resources: `http://dmp.fairdata.solutions/resources/jxb`

2.6   IS THE DATA RESOURCE VERSIONED?

What's up?

Core reference resources especially, are more often than not versioned (updated).

Static (OPEDAS) datasets from individual experiments may or may not be updated but can be changed over time. In any case, the influence of (reanalysis) of your data that changing OPEDAS sets or workflows may have should be considered in the study planning phase, and appropriate measures should be taken to avoid undue surprises when re-using the 'same' OPEDAS resource at a later date. Questions to ask when analysing OPEDAS resources before reuse include:

If the source is updated, will you do your analysis again? What level of detail/granularity of the data do you need and will that be affected by updates? If you need only part of the data, can you filter it before downloading, and what is the subset you really need?

DO

- Check the versioning policies of the OPEDAS source and consider the consequences.

- Decide what version to use.

- Decide what you will do when updates are released.

- In case you always want to use a given version, make sure you will always have access to that version.

- This means that if the source does not freeze versions, you may have to download, store, and document the version you will use.

- Make sure you extensively record and publish which version you used in which analysis.

- If possible, subscribe to updates of the resource so that you will be aware of updates.

- If no update policy is clearly described at the source, try to find out the actual situation.

- Propose to the owner or custodian of the resource to add information on the update policy.

DON'T

- Use versioned OPEDAS sources without recording the version and installing a routine for updates.

- Assume that the OPEDAS owner/repository does have a proper versioning and documentation policy, unless this is explicitly stated and described.

- Publish results of your analysis without referring to the exact version of the OPEDAS resources you used for analysis.

- Make any claims in your research output that are based on OPE-DAS, which may render the reference to the OPEDAS obsolete and/or confusing to the users of your research output.

Resources: http://dmp.fairdata.solutions/resources/rgy

2.7 WILL YOU BE USING ANY EXISTING (NON-REFERENCE) DATASETS?

What's up?

In case some of the OPEDAS sources you want to use are not qualified as 'reference datasets', and/or are not available in TDR-type repositories where access, ownership, sustainability, and versioning are well documented, a number of extra questions need to be answered before you can responsibly reuse the data for your analysis.

DO

- Check the conditions under which you can get access.

- Check ownership.

- Check potential restrictions (for example, commercial use, which might come in much later).

- Decide what version to use (if versioned).

- Double check with the owners as to how long they intend or guarantee to keep the data available in the same format and version.

DON'T

- Expect that data not explicitly deposited in a trustworthy, sustainable environment will be there when you next visit.

- Use OPEDAS indiscriminately, without detailed recording of their origin.

- Expect your data analysis workflows to give reproducible results unless you guarantee stable input, including the OPEDAS-dependent elements of your analysis pipeline.

Resources: `http://dmp.fairdata.solutions/resources/wya`

2.8 WILL OWNERS OF THAT DATA WORK WITH YOU ON THIS STUDY?

What's up?

Some OPEDAS may not be usable without explicit consent of the owner and without assistance of the owner. This does not mean the data are necessarily badly documented or of low quality. They may be restricted by privacy laws or connected to a tissue biobank or even personal data of an individual citizen who is the legal owner of a personal data locker. If you need cooperation from an OPEDAS owner, there are additional issues to consider, such as time constraints, co-authorship, and informed consent.

DO

- Check the conditions under which you can get access, and contact the owner.

- Prepare a clear explanation of why you want to reuse the data, and for what purpose.

- Ensure the level of explanation fits fhe owner.

- Check verbally and in writing with the owner about potential restrictions that might not be explicit in the metadata of the OPEDAS.

- Double check with the owners as to how long they intend or guarantee to keep the data available in the same format and version.

propose

DON'T

- Expect that all OPEDAS owners are prepared to share their data for any research purpose.

- Underestimate the burden you may put on the OPEDAS owner.

- Ignore other incentives than 'citation', such as, for instance, a formal acknowledgement or co-authorship.

- Ever use data you find on the web or elsewhere without making absolutely sure you are not violating informed consent or other generic rules.

Resources: `http://dmp.fairdata.solutions/resources/dcy`

2.9   IS RECONSENT NEEDED?

What's up?

In case OPEDAS are subject to informed consent rules, you have to ensure that the consent given covers the specific purpose of your intended study.

The GDPR[1] in Europe and the equivalents in other regions need to be respected. Even if you do not expect it, re-consent of the OPEDAS owner(s) may be a legal prerequisite (even the owners may not be aware of this legal situation), and therefore, utmost care has to be taken that you do not violate any laws. It is also advisable to discuss the reuse of OPEDAS with your ethical committee in case it involves data directly derived from studies involving animal or human subjects, or data from, for instance, 'social media' origins.

---

[1] https://www.eugdpr.org/

DO

- Check the consensus statements associated with the data (if any).

- Contact the data owner actively when there is any doubt about the legal issues associated with the reuse of the data.

- Prepare a clear explanation of why you want to reuse the data and for what purpose.

- Ensure the level of explanation fits the owner.

- Double check with the owners as to how long they intend or guarantee to keep the data available in the same format and version.

DON'T

- Expect that all OPEDAS owners are prepared to share their data for any research purpose.

- Underestimate the burden you may put on the OPEDAS owner.

- Ignore other incentives than 'citation', such as, for instance, a formal acknowledgement or co-authorship.

- Use any potentially sensitive or otherwise restricted data for your experiments without all the checks above, even if they 'seem' to be freely available on the Web.

- Take for granted that data that can be used for pre-competitive research can also be used for commercial purposes (when you work for a company, license and consent issues are even more pressing in some cases than for public research institutions).

Resources: `http://dmp.fairdata.solutions/resources/bqy`

2.10   DO YOU NEED TO HARMONIZE DIFFERENT SOURCES OF OPEDAS?

What's up?

Once you have decided which OPEDAS resources you wish to use for your study, there are several questions to be asked regarding different data formats, harmonisation issues, and, for instance, the use of identifiers and terminology used in the data. If these questions are not addressed up front, they may severely delay or even completely jeopardise the study.

DO

- Check the format in which the actual data elements are presented.

- Check the proper terminology systems used for each dataset and discuss consequences for analysis pipelines, etc.

- Decide how the formats of the OPEDAS resources and the ontologies used will influence your choices regarding your own data capture.

- Check whether metadata or data elements contain natural language fields, or if they need translation or mapping to ontologies

- Where possible, consider reformatting data to make them 'linkable', and report back to OPEDAS owner.

DON'T

- Expect data (even if in FAIR format) to be reusable in your setting without any reformatting.

- Compare 'apples and oranges', because data are not properly harmonised and might give the wrong correlations.

- Change anything in OPEDAS sets (formats, translation of terms, mapping to term systems) without proper documentation, and provenance, and report these to the OPEDAS owner.

Resources: `http://dmp.fairdata.solutions/resources/wht`

2.11 WHAT/HOW/WHO WILL INTEGRATE EXISTING DATA?

The nature and size of the selected OPEDAS for your study will determine to a great extent how much work it will be to get all OPEDAS sets accessible (online or locally) in the format required for your analytical procedures. Here we help you consider some questions that can guide you towards a good integration and analytics plan and assist you in properly budgeting for the required activities.

2.11.1   Will you need to add data from the literature?

What's up?

In case you consider using (machine-readable) data that are directly mined from the literature, you should carefully consider whether you should mine these yourself or whether appropriate machine-readable collections of the relevant literature already exist in the machine-readable format. Many groups have specialised in text-mining, and some of them provide the results of their decade-long efforts in FAIR format. Unless you are in a specialised text-mining group, chances are that it will cost you a prohibitive effort to reach anywhere near the same quality as these specialised groups can offer.

DO

- Check whether collections of pre-mined data from the literature are available in machine-readable/analysable format.

- Treat these as OPEDAS and contact the OPEDAS set owner (usually the text-mining expert) to check on disambiguation levels, precision and recall, and other key features of the dataset.

- Ensure that the output data are in the correct format, and consider the balance between, for instance, mapping the data to a different term system versus re-mining them with a different tagger (the latter may actually be easier and faster if you collaborate with the experts).

DON'T

- Think that computers can properly read text, they will give you very ambiguous and noisy results, due to homonym, synonym, and syntax problems.

- Consider using text-mined results without studying in some depth the many issues associated with conversion of human- readable narrative to machine-readable data.

- Try text-mining yourself with amateur tools as it will almost surely lead you into a time-consuming and unproductive sidetrack, contact experts instead and collaborate.

Resources: `http://dmp.fairdata.solutions/resources/pth`

2.11.2   Will you need text-mining?

What's up?

text-mining (recovering structured information from unstructured text) is a discipline in itself. You will find some literature here so you can read up on it if you need to. There was a time when major textual resources were not mined properly, and individual researchers needed to recover concepts and their relations from text. However, nowadays, for many text corpora, there are well-designed and state-of-the-art collections of concepts, co-occurrences, and relations mined by specialised consortia. Even these specialised groups will only seldom reach the 80/80 level of precision and recall, but they are probably the best you are going to get. There may obviously be internal or obscure texts or textual collections that are not mined yet, and therefore, if the information source is crucial to your research, you might consider including text-mining steps in your own protocols. However, again be aware that this is a very complex field, and you may end up working on disambiguation, machine learning, thesauri, concept taggers, and the like instead of doing your research.

DO

- Double check whether the text corpus you are about to mine has not yet been mined and the results made available by others (see external resources).

- Ensure that text-mining is unavoidable. If the associations you hope to find are in public databases already as a result of previous text-mining and/or manual curation, you will most likely be wasting your time to try and do better.

- Consider outsourcing the text-mining (if really needed) to a specialised group or company. They are likely to get much better results than you will get with a 'home-made' mining algorithm.

- Use 'ready-to-use' text-mining software if you can not or will not outsource the task.

- Make sure that the mapping of terms to concepts in your output is correct and compatible with the data which with you want to combine the output.

- Budget for text-mining in your research plan if it is unavoidable as a workflow step.

DON'T

- Underestimate the complexity of text-mining as a method.

- Mine concept or relationships from any text without considering all other data stewardship issues as they are described here for any other data type (text-mining results are highly about intensive and therefore potentially valuable data).

- Develop new text-mining algorithms unless your team agrees that none of the (many) existing systems are suitable for your purposes.

Resources: `http://dmp.fairdata.solutions/resources/jyd`

2.11.3   Do you need to integrate or link to a different type of data?

What's up?

The collection of OPEDAS sets you may want to exploit during different phases of your new study will partly determine how you go about setting up the actual experiments, and, also in particular, how you capture and format your own data. If the majority of the OPEDAS sets you need are in a given format, mapped to a particular set of term systems, and can be processed by a given range of workflows, this may drive you towards generating your new data in a format and semantic environment that are as close as possible to the features of the relevant OPEDAS sets.

DO

- Check what the predominant data formats are, and make a list.

- Make a list of term systems used by the OPEDAS owners for concept-mapping.

- Discuss as a team how these lists might influence choices to be made during the next steps in the data cycle, in particular, your data capture, term systems and standards used, and formatting.

- Check the compatibility of all datasets with the intended work-flows to use for data analytics, and determine the work needed to harmonise and reformat particular OPEDAS datasets for the analytics pipeline you anticipate using.

- Use all this input to determine in detail what is needed to enable you to start an integrated analysis of the OPEDAS sets (even if you do not generate new data, this may still be a lot of work).

- Make a detailed data re-formatting and capture plan geared towards the data analytics needs you have in mind.

- Budget for that in your research plan.

DON'T

- Assume that workflows you choose for your analytics will consume and combine different data formats without any prior harmonisation.

- Capture data first in a format that you happen to know, and then risk finding out that you should have captured your data in a different format, with a different term system to map to and/or with different granularity.

Resources: `http://dmp.fairdata.solutions/resources/ajm`

2.12 WILL REFERENCE DATA BE CREATED?

What's up?

If you create data that are meant to become a reference dataset, there are even more challenging details to be aware of. In addition to all data stewardship considerations described for 'normal' data, the creation of reference datasets is very specifically directed at intended reuse. Just as with software, when other people try to use your code, lousy documentation is very bad practice, and causes other people a lot of trouble. Also please note that creating reference data is not necessarily an act of experimental data creation. It is very likely (if not the rule) that reference data are created by collection, reformatting, integration, curation, and annotation of existing data. After these cumbersome data-munging processes, the resulting set is then offered for reuse by others. Some such databases (like UniProt) became so-called core resources: Almost everyone in the domain uses them and will reference them in new studies. So, citability, professional versioning of the data itself, and, for instance, APIs become even more critical than with individual new datasets.

DO

- Check thoroughly whether reference data already exist (creating new reference data is very expensive and time consuming).

- Check what the predominant existing community formats and standards are for the data type(s) you intend to create.

- Make a list of term systems to be used for the concepts to be referred to in your database.

- Give construction of the data infrastructure, performance of your ICT infrastructure, and your ability to maintain and update the data (if needed) sufficient attention.

- Consider early on, working with larger and professional data centres that can set up, maintain, and support the reference data.

- Carefully budget for the creation and maintenance of your reference data in your research plan.

DON'T

- Create reference data without a very specific scientific purpose and need.

- Create redundant reference data.

- Host reference data in your own academic research environment/infrastructure, if at all possible.

- Assume that 'others will take over' your reference data and maintain it properly without first checking that this is the case.

- Create reference data without a proper release, update, API, and versioning policy plan.

- Create data first and then realise they could or should be reference data, and now, you did not follow the correct procedures.

Resources: `http://dmp.fairdata.solutions/resources/rbz`

2.12.1 What will the IP be like?

What's up?

For new data you created, the IP is *a priori* with 'you' or your organisation, unless otherwise specified by the funding body supporting your research. For data on human subjects, however, the legal ownership of the data is most likely with the individuals that participated in the study. In that case, you are a 'custodian' of other people's data by default. This brings many additional responsibilities to you as a data steward. Obviously, considering how to deal with that IP is always important, also for *ad hoc* experimental data, initially created for internal use. However, for the creation of reference data, and specifically when it means the extraction, transformation, reloading, curation, and 're-packaging' of OPEDAS to create a reference dataset, the IP situations and the license under which these source data can be reused are extremely important. The legislation in such accumulated databases (including their implicit ownership) can be different in different countries and regions. The licenses of the source data may 'carry over' to your new dataset and include elements with highly restrictive licenses (for instance, restricted to non-commercial use only) that may later jeopardise your attempts to make the reference dataset widely used, and your ability to 'exploit' the reference data for sustainability purposes.

DO

- Thoroughly check the IP situation and the licenses of all data sources you intend to use for inclusion.

- Also do this if you only include small subsets of larger resources, even if they are 'public' at first sight and not password protected (there might still be restrictions).

- Check whether the sources you want to include have properly dealt with the IP situation of sources they included.

- Contact data source owners when you have doubts about the licensing or IP situation of that resource.

- Consult legal and licensing experts to ensure that you do not violate any explicit or implicit rules by creating the intended reference dataset.

DON'T

- Create reference data without FIRST consulting a licensing expert.

- Create reference data without doing all these checks, because you think 'only your own department will use them' (that would almost disqualify them as 'reference data' in the first place).

- Underestimate the time and effort needed and the scrutiny of, for instance, whether the pharmaceutical industry uses your reference data (and maybe even pay for it) later. Even datasets without any license (and seemingly open) form a strong liability for large commercial companies to sue, in formal approval processes for instance.

- Choose a license for your own data without professional advice and use the most open license acceptable for your reference data (in Open Science number of users roughly equals chances for sustainability).

Resources: `http://dmp.fairdata.solutions/resources/hct`

2.12.2   How will you maintain it?

What's up?

Reference datasets can be very important for interpretation, review, and reproducibility of other people's experiments and results. Creating reference data is therefore a contribution to the common good of research data and, therefore, long-term maintenance is a key issue. A frequent problem is that reference datasets are initially set up at a rather small scale, within specialised academic institutes. Also, sustainability is only addressed towards the end of the research project that generated these datasets. Short-cycle funding systems are, generally speaking, not sufficient to guarantee long-term sustainability of reference (core) resources. It is therefore of great importance to plan the creation and the long-term maintenance of reference data in advance, and to carefully consider choices that will influence maintainability later on, such as data format, level of automation versus manual curation, data infrastructure choices, number of expected users, network capacity, and the ability and or willingness of your institution to maintain 'public resources' beyond the immediate use for internal research purposes. If any of these are 'negative' your resource may actually bring other people that use it into trouble in the long run, as part of their scientific discourse relied on your data, and non-availability will affect them directly.

DO

- Check thoroughly with your supervisors and institute whether maintenance of reference data beyond the research project is an option.

- Seek advice from professional data owners (preferably of larger. sets than the one you intend to create) to ensure optimal planning of the supporting infrastructure and the curation and maintenance effort. Reuse of your data by others may bring with it significant correspondence, support effort, but also co-publications, and discuss this with your supervisor.

- Specifically discuss a maintenance plan with your group, support staff, and supervisors, including a versioning and release plan (for both the data itself and the application programming interfaces (APIs) and graphical user interfaces (GUIs) if relevant) and a customer-support plan.

- Consult legal and licensing experts to ensure that offering the dataset is maintainable, and reusable from a legal perspective.

- Preferably work with a professional (sometimes commercial) data provider for the 'commodity' part of maintaining the reference data resource, and separate the 'research and innovation' environment very strictly from the 'commodity part' both in terms of funding approach and support.

DON'T

- Create (potential) reference data with public money with the intent to close them off for other researchers unless there is a very clear and defensible reason for this (you will increasingly have to argue very strongly for non-open data in your research proposals and keeping data closed may lower your selection chances).

- Assume that 'someone in the department' will pick up the longer-term stewardship of your data because they are 'so obviously valuable'. Many universities have no incentives to structurally 'provide services' to third parties.

- Create reference data without the utmost care to make both the data elements themselves and the metadata FAIR and machine actionable wherever possible.

Resources: http://dmp.fairdata.solutions/resources/usx

2.13 WILL YOU BE STORING PHYSICAL SAMPLES?

What's up?

Let's define a 'sample collection' as any collection that contains physical (reference) samples. Obviously, most sample collections, such as biobanks, are collected for research purposes. In case the samples are of human origin, or otherwise associated with ethical and privacy considerations for any reason, the stewardship aspects are significant. As you may have noticed, the term 'data' was not coupled with the term 'stewardship' here. We can argue that even a tissue sample is a 'form of data' but, as it is not digital we will separate the discussion about the physical samples from the data stewardship associated with biobanks. The biobanking world is better organised than many other fields, and therefore all the issues related to the stewardship of (samples in) biobanks or other sample collections will not be covered here. It should be emphasized, however, that the annotations and the metadata of the samples and the sample collection as a whole should be FAIR like those of any other data.

It is also clear that in many cases, broad metadata about a biobank and the broad collections it contains can be 'open', while the actual details on and access to the samples themselves are highly controlled and restricted, for obvious reasons.

DO

- Check thoroughly with experts in your institution to find out whether biobanks and all related infrastructures already exist in your institution.

- Seek advice from professional biobank owners (most likely of larger sample and datasets than the one you intend to create) to ensure optimal planning of the supporting infrastructure and the curation and maintenance effort. (Reuse of your samples and data by others may bring with it significant correspondence, and support effort, but also co-publications; discuss this with your supervisor.)

- Specifically discuss a collection maintenance plan with your group, support staff, and supervisors, including a versioning and release plan (for both the samples and the annotations and metadata) and a customer-support plan.

- Consult legal and licensing experts to ensure that offering the samples and dataset is maintainable, and reusable from a legal perspective, including the options to deal with informed consent.

- Preferably work with a professional (sometimes commercial) infrastructure and data provider for the 'commodity' part of maintaining the reference samples and data resource, and separate the 'research and innovation' environment very strictly from the 'commodity part' both in terms of funding approach and support.

- Make sure you have a strong sustainability plan.

DON'T

- Create a sample collection based on public money with the intent to close it off for other researchers unless there is a very clear and defensible reason for this. You will increasingly have to argue very strongly for non-open collections, and especially their metadata in your research proposals, and keeping data closed may lower your selection chances.

- Assume that 'someone in the department' will pick up the longer-term stewardship of your samples and data because they are 'so obviously valuable'. Many universities have no incentives to structurally 'provide services' to third parties.

- Create reference samples and data without the utmost care to make both the annotation elements themselves and the broader metadata FAIR and machine-actionable wherever possible.

Resources: `http://dmp.fairdata.solutions/resources/kuz`

2.13.1   Where will information about samples be stored?

What's up?

Let's define two different kinds of data (in nature, not in format) about samples or other physical objects in a research collection that cannot themselves be made FAIR in the sense of intrinsically findable, accessible, interoperable and reusable, including by machines. This holds true for 'tissue' or other physical samples, but also, for instance, for non-digital images like paintings, or a natural history collection of specimens, biological samples, or you name it. These 'research objects' cannot be made FAIR themselves, but their 'descriptions' (annotations) starting with the legends on a video or a picture, can grow to full annotations and 'metadata' on entire collections. Although one could argue that anything that is 'asserted' about a 'sample' is a form of metadata and is also considered an 'annotation', we will still make a distinction. We defined annotations in this context as a subset of 'user defined' metadata (see Introduction) that give a functional description of the sample and its characteristics, what it was meant for, what it has been, and can be used for, etc. More classical metadata are the intrinsic metadata about the sample itself, such as methods, time stamp, location, and metadata about increasingly aggregated levels of the collection (up to the address of the building where the specimens are located).

DO

- Apply all criteria for both annotations and intrinsic metadata as described for reference data in general and register your collection in one of the existing catalogues.

DON'T

- Create a sample collection based on public money without proper FAIR annotations and metadata. Even for your own research, this is crucial, but sharing your data as far as is legally allowed is the default and without FAIR metadata this will be seriously impaired.

Resources: `http://dmp.fairdata.solutions/resources/fqu`

2.13.2 Will your data and samples be added to an existing collection?

What's up?

It might be necessary to start an entirely new collection or biobank of research materials for your study, but in many cases it might be much more efficient and sustainable to add your samples and the associated annotation and metadata to existing environments. Reasons to start an entirely new collection might be that the samples and the data are so sensitive that they cannot legally leave your institution, while the institution does not yet have any 'biobanks'. In that case, you need to first think carefully as to whether the collection will be sustainable in the first place. What happens to the data and the samples after your study is finished if no one else in your institution ever saw the need, so far, to create sustainable research collections in your institute? This will be a question reviewers will increasingly ask in data stewardship sections of research proposals. So, as argued before, first go through a careful process of finding out about existing biobanking initiatives in your institution or research consortium. If you decide that you have solid and defensible reasons to start an entirely new collection rather than adding a sub-collection to an existing infrastructure, try to use generic criteria and infrastructure developed for professional biobanking in the relevant national and international consortia in this field.

DO

- Go through a very serious (and documented!) effort to search for opportunities to optimally use existing research sample collection infrastructure in your institution or consortium.

- Make sure the existing collection or your new one applies all criteria for both safeguarding or sample quality and preservation as well as the FAIR principles, for all annotations and intrinsic metadata, as described earlier for reference data in general, and register your collection in one of the existing catalogues.

DON'T

- Create an isolated and likely unsustainable biobank based on public money without proper infrastructure and without FAIR annotations and metadata. Even for your own research, this is crucial, but sharing your data as far as is legally allowed is the default, and without FAIR metadata this will be seriously impaired.

Resources: `http://dmp.fairdata.solutions/resources/hhg`

2.14   WILL YOU BE COLLECTING EXPERIMENTAL DATA?

What's up?

Please note that it is no longer self-evident that answering a given research question needs *de novo* experimental data. More and more examples appear in the literature that demonstrate how new discoveries can be made entirely based on 'OPEDAS'. As the generation of new experimental data is very expensive and time consuming, the rigorous questioning of the need to do so is part and parcel of good data stewardship for discovery. For a data-driven scientist, the optimal use of OPEDAS should be a default attitude. In many cases, however, you may need to create a (usually comparatively small) dataset based on experiments, questionnaires, or other observations. In that case it is obviously crucial that your data 'talk' to OPEDAS relevant to the interpretation of your new data, and that they can be processed by machines and workflows in conjunction with OPEDAS. Therefore, in addition to the usual statistical and 'methods' considerations when designing an experiment or a study, the way data are captured, formatted, and published is a crucial data stewardship consideration, and therefore, data stewards should be regarded as being essential expert colleagues already in the design phase of any study. Again, data stewardship is not just a service to others for whom your data will become OPEDAS, but the analysis and interpretation of the new data will be infinitely more easy, effective, and reproducible when they 'feed into' standardized formats and workflows.

DO

- Go through a very serious (and documented!) effort to search for opportunities to optimally use existing research data collection infrastructure in your institution or consortium.

- Make sure your new data collection applies all criteria for both safeguarding or data quality and preservation as well as the FAIR principles for all data if possible, but certainly for all annotations and intrinsic metadata as described for reference data in general, and register your collection in one of the existing data catalogues.

DON'T

- Assume that your data will never be used as OPEDAS or even reference data, and therefore, that 'personalised data steward-ship' (a metaphor for messing with your data in isolation) is acceptable.

- Create an isolated and likely unsustainable dataset based on pub-lic money without proper infrastructure and without FAIR anno-tations and metadata. Even for your own research, this is crucial, but sharing your data as far as is legally allowed is the default, and without FAIR metadata, this will be seriously impaired.

- Create more (or less) data than can be reasonably defined to be necessary for your study and without proper statistical and data analysis advice from the experts in your institution, or (if not available) in your research consortium or elsewhere.

Resources: `http://dmp.fairdata.solutions/resources/csx`

2.15 ARE THERE DATA FORMATTING CONSIDERATIONS?

Raw data (now increasingly created by machines) come in a wide va-riety of formats, some of them even being proprietary to the vendor. However, in many cases, the actual 'analysis' of your research results will only start after some sort of 'pre-processing', of the raw data in a format that is a suite of the analysis approaches and pipelines you want to run. It is obviously critical to keep rich provenance informa-tion about 'how you did this' and a strong and persistent link to the raw data files. However, the real data stewardship challenge started when you pre-processed the data into other formats, rather than with what was 'spit out by the machine'. Raw data can range from pictures, graphical outputs, and tables to sequence reads and, in extreme cases, just strings of 'zeros and ones'. It is obvious that many of these raw data files cannot (and need not be) FAIR, as we will effectively work with the pre-processed derivatives of the raw data.

Data processing may be error prone, and therefore it should always be possible for people reusing your data later, to access the raw data files if they are stored, which is not always the case. If you delete certain raw data (like images from a DNA sequencer that precede the actual sequence output), that fact should also be part of your provenance.

The choice of the data format in which you express your preprocessed data is crucial for future use and reuse. Proprietary and 'local' formats, identifiers for concepts, etc., may severely impair future reuse. In some cases, this may be deliberate for privacy or competitive reasons, but we will work here under the assumption that your data stewardship plan includes the option (and goal) of making your data FAIR for others to reuse later.

For almost any type of data that are 'frequently created', there are community-emergent data formats that are more popular than others. It is therefore a critical step to choose the format carefully. The intrinsic size and nature of your data is an important factor, but so are the workflows you want to run on them. Here again, if commonly used and supported workflows exist, these are preferred. The formats that such 'established' workflows support would have preference as a more sustainable choice. So, in general, only develop or choose home-made workflows and data formats if a thorough search has not revealed any suitable existing options.

2.15.1 What is the volume of the anticipated dataset?

What's up?

Volume is only one of the aspects of your dataset (others are complexity, privacy, variety, etc). However, the volume may bring some formatting considerations of its own. In some cases you may be able to store only part of the data without the loss of essential information. For instance, if you have done full genome sequencing on a group of individuals, you may be able to store one reference genome (with rich provenance, of course) and store only the sequence variations of each individual distinct from that reference genome. This will reduce your stored dataset enormously (less than 1% of the original sequence), and at any given moment you may be able to 'regenerate' the entire genome of each individual by adding the rest of the sequence from the reference genome file.

DO

- Always consider data volumes first and decide how these reflect on choices you make in preprocessing choices, provenance storage (very detailed logging of what happened to the original raw data, how, when and why.

- Format of the processed data, based also on expected workflow choice, and common practices for the data type.

- Check, prior to collection, that your institutional (or external) storage and compute infrastructure and access is sufficient to properly preprocess, store, and access the data.

- Anticipate as well as possible your own groups' current and potential future use of the data.

DON'T

- Assume without checking that the compute and storage facilities you have access to are adequate.

- Create any file formats, vocabularies, or other data-related assets without first convincingly demonstrating that no reusable solutions exist.

Resources: `http://dmp.fairdata.solutions/resources/gqs`

2.15.2   What data formats do the instruments yield?

What's up?

Most instruments in laboratories put out data in one or multiple formats that are specified by the manufacturer. In many cases, the downstream processes of data pre-processing, transformation, curation, and linking will transform these data into other formats. It is very important to be fully aware of all data formats and standards that are 'machine' or 'instrument' imposed before the actual data capture and downstream processes begin.

DO

- Make sure you know all instruments to be exploited in the planned study and their imposed formats and limitations.

- Make the data format that is the standard output of any instrument part of the provenance trail and metadata.

- Record all relevant information about instruments.

DON'T

- Start measurements without the instrument landscape being fully clear, including constraints and limitations imposed by them.

- See the instrument part of the research as the sole responsibility of the experimental researcher and just passively wait for the data to come out.

Resources: `http://dmp.fairdata.solutions/resources/yqk`

2.15.3 What preprocessing is needed?

What's up?

As said before, in many cases you will process data in some way to prepare them for further analysis.

The preprocessing of data is sometimes standard and very easy, but it can also be almost a scientific challenge in and of itself, especially when instruments and approaches are relatively new.

Deep understanding of the scientific goals, the context of the data, their structure, and the underlying assumptions is needed to make the right choices concerning the processing. Here, we cannot go into detail about the hundreds of data/processing combinations that may need to be considered, but suffice it to say: this is a matter of intense study with content and data specialists involved. Also, the provenance (logging, history) of precisely what happened to the data after their initial creation is absolutely critical.

DO

- Ensure that full understanding is reached between content specialists and data specialists about the issues mentioned above before action is taken.

- Carefully consider what the data represent and how they will be used downstream in your own experiments and potentially later by others.

- Log any processing you did to the raw data very precisely and in a FAIR format.

- Make a permalink between the processed data files and whatever raw data may be preserved.

DON'T

- Assume that you know how to do the processing and and that the processed data is all that matters. Increasingly, (data) publishers and funders will require the option to 'return to the raw data'.

- Store raw data without FAIR metadata and in a repository that is less trusted than where you store the preprocessed data. (Note: These may be very different, however, for instance, you may need to store raw data on tape, while the processed data may have to function in high performance and re-analytics environments and be readily accessible to external workflow systems.

- Store useless raw data (good data stewardship also means deleting data that cannot conceivably be of any further use).

- Assume too easily that raw data are useless and fail to think about the long-term future, other users, and other disciplines.

Resources: http://dmp.fairdata.solutions/resources/fqv

2.15.3.1   Are there ready-to-use workflows?

What's up?

For many kinds of data, there are established and well-tested workflows and pipelines, turning raw data into analytics-ready formats. Make sure you are fully aware of these options for the kind of data at hand. Reproducibility of results and conclusions is strongly related to the standardisation of data formats, and to the quality and the robustness (and version) of the workflows you use. The documentation and versioning of academic workflows (professorware) is not always up to standard, and using the 'wrong version' of a workflow and its components later may give significantly different results on the same data used as a substrate. In some cases, 'custom' analytics workflows are unavoidable, but if you have to use 'experimental' workflows, take extra care to document exactly what version you use.

DO

- Study tool and workflow registries to make sure you use established and reproducible workflows wherever possible.

- If these do not exist and prototypic workflows need to be used, or when you have to develop custom workflows, make great effort to document code, and versions of tools used (like the version of a vocabulary used for mapping).

- Properly archive code, versions, and components used, and document then as richly as possible.

- Add FAIR metadata to your workflows so that they could be found, accessed, operated and reused by yourself and others with maximum range of reproducibility.

DON'T

- Use custom-made workflows unless absolutely unavoidable.

- Run these workflows without extensive documentation, assuming that you will be the only one to process the data and only once.

- Store workflow code and components on your local system without proper backup and metadata.

Resources: `http://dmp.fairdata.solutions/resources/eif`

2.15.3.2   What compute is needed?

What's up?

One of the most frequent mistakes is the assumption that sufficient compute will be available either locally or externally (service providers and cloud solutions). First of all, the ever-expanding use of powerful high-throughput technologies may generate datasets of a size and complexity that cannot be properly handled by the local systems available to you (and you may have to book capacity). Storage is one thing you must have checked, but compute needed to process and analyse the data is yet another matter. Some data might be too large or too privacy sensitive to 'ship outside the firewall' of your institution. In that case the compute needs to go to the data inside your institution, and sufficient compute needs to be locally available to run all processes needed. In many cases, when you come to the analysis phase you will need to use or download workflows and data from other places. Therefore, compute capacity and expertise in your institution must be able to cover all these needs, and these need to be specified and checked before an expensive and complicated experiment is started.

DO

- Make sure you know the ICT department that will have to 'deal with your data' if you do not control the hardware yourself.

- Plan carefully with them how (and when) your data will be delivered to them and what processes are expected to be run.

- Work with the compute specialists in your department with great respect, and discuss needs and also particularly compute burden as well as (long-term) storage consequences of your data.

DON'T

- Ever underestimate the complexities related to professional data processing, and the needed compute and storage infrastructure, skills, and costs. If you find out later that you underestimated this, you will run into troubles, annoy your expert colleagues and you may even lose your data.

- See the 'ICT department' as a necessary but less interesting and 'difficult' part of the institution, and treat your colleagues as your 'data handlers' only, but engage them where possible in your research.

- Note: This is a typical place where 'local expertise and resources' as a 'button in the website' can be customised for the institution. Existing collections stewards can actively keep records here of what exists, thereby minimising the chance that naive PhD students and their equally naive (or overcommitted) PIs will unnecessarily start from scratch.

Resources: `http://dmp.fairdata.solutions/resources/qej`

2.15.4 Will you create images?

What's up?

Images pose particular challenges. Especially those collected with modern, high-resolution instruments, and frequently sequentially collected. Such approaches generate very large datasets and high costs in terms of storage. They also contain a lot of 'intrinsic information' that makes them interpretable by people (and in some cases machines) probably even beyond your (current) imagination. Also, images are much easier to interpret when viewed in context, and therefore, if you need to anonymise or pseudonymise pictures (for instance, medical images), make sure that you do not preclude that process during image creation or processing. The most important issue here is to separate annotations and metadata from the actual image, so that machines do not get confused between the image itself and the annotations.

DO

- Consult colleagues on the need to be able to deliver the images to researchers (internal or external) without annotations and metadata embedded.

- Also consider this when the images do not have privacy issues attached; there may be scientific reasons to preserve the ability to use and share the images without embedded annotations and metadata (for example, for machine-learning purposes).

- Adorn pictures with rich and FAIR annotations and metadata (both intrinsic and user defined).

DON'T

- Embed annotations or metadata in the picture itself. This may seem logical, as it will make them 'inseparable' from the image itself, but in actual fact it may severely restrict the reuse of the images for purposes other than your original intended use.

- Store pictures without a very strong link to the associated metadata and annotations, but make sure you can always separate the two when needed.

- Treat metadata and annotations of images with any less care than generically described for datasets.

Resources: `http://dmp.fairdata.solutions/resources/mkg`

2.16 ARE THERE POTENTIAL ISSUES REGARDING DATA OWN-ERSHIP AND ACCESS CONTROL?

What's up?

There is a clear difference between legal data ownership (usually with the creator of the data or those who funded the experiments) and the degree to which the data will be made 'open' to and shared with others. Even entirely 'open' data have a legal owner. As insights into the extent to which data will be shared may change over time, initially you need to treat all data as 'private' to the creators. Opening up data is always possible, but keeping data restricted is sometimes no longer possible when the wrong choices have been made during creation, processing, or publishing and licensing of the data.

DO

- Consider data ownership and intended use, also specifically beyond the project in which the data were generated, as early on in the process as possible.

- Discuss ownership and control issues with the relevant experts in your institution (or external experts).

- Choose a license for the data when they are ready to be published and consider reuse options, including citation requirements.

DON'T

- Create data first and worry about ownership and access later. Features of the data may restrict choices later on in the process.

- Ever consider data as 'just for this experiment' and thus underestimate reuse issues. Even if the data will be never reused in further integrated analyses, they still need to be always available (also to others like reviewers or auditors) for reproducibility checks.

Resources: `http://dmp.fairdata.solutions/resources/acx`

2.16.1 Who needs access?

What's up?

There is a tendency to treat data as being for the sole purpose of the experiment for which they were generated. Even if that is the case, there will usually be multiple people in the team (and not always only in your own institution) that need regular access to them. Consider not only the legal, privacy, and other 'social' issues around the data, but also the technical hurdles that may arise when the data must be shared within the research group or consortium. This is not only related to the data themselves. If you have, for instance, used proprietary tools to preprocess the data, or tools that are free for academic use but restricted for commercial use, you may not be allowed to share them with private, commercial partners in your research consortium. Also, if data cannot be moved due to size or sensitivity, make sure that all partners that will need them will be able to access them locally.

DO

- Discuss intended data generation with all your consortium partners very early in the process.

- Ensure that licensed tools and, for instance, vocabularies that are used to process, map, or format the data allow the sharing of the data in the entire consortium.

- Consider the size and the complexity of the data, and anticipate any hurdles that these features may imply for the ease of sharing the data.

DON'T

- Assume that all open source tools and formats do allow all kinds of sharing.

- Assume that all partners in the consortium only have academic and pre-competitive needs concerning the data.

- Assume that all partners have sufficient bandwidth, local storage, and compute to deal with the data, without checking first.

Resources: `http://dmp.fairdata.solutions/resources/cvk`

2.16.2 What level of data protection is needed?

What's up?

Whenever data are generated, privacy and security issues need to be considered. This does not only hold true for personal data on people. In many cases, the institution will have a policy about keeping data 'internal' and thus restricted in access. The reasons may be legitimate even in publicly funded research, where the default usually is to make the data ultimately open (or rather FAIR). In many cases, data processing, quality control, curation, and interpretation are lengthy processes and at least during these processes it can be counter-productive to give others than the data owners access to the data. They are simply not ready for reuse by others yet. So even if the data are intended to be made entirely public at the end of the pipeline, they may have to be restricted in access until the experimental evaluation is finished and the data are ready to be published.

DO

- Check with all investigators whether temporary or structural protection of the data is needed.

- Estimate the period over which the data needs to be protected, and check the availability of infrastructure, funds, and procedures to enable that.

- Estimate the costs of long-term protection of the data after the experiments have been concluded and the results published.

DON'T

- Assume that data (even if there are no privacy and sensitivity issues) can just be put on 'any laptop' in the institution. You might still be violating internal procedures.

Resources: `http://dmp.fairdata.solutions/resources/ajz`

2.16.2.1 Is the collected data privacy sensitive?

What's up?

If data (for instance, on patients) need to be structurally protected, you need to go through an in-depth evaluation of the levels of security needed, anonymisation, pseudonymisation, and/or encryption of the data. You also need to make sure the consent obtained from the patient does allow all the studies you want to perform on the data. An increasing challenge is that with increased reuse of valuable datasets and cohorts, reuse beyond the original consent, is more and more frequent. Without renewed consent these further intended experiments may be illegal and therefore you need to obtain the broadest possible consent that is allowed by the ethical committees and acceptable for the individual patients. One option is to keep control of the personal data entirely in the hands of the research subjects, so that they can give informed consent for every future study that may want to involve their data. In that case, special technologies are needed (see PHT). In all such cases, you need to be extra aware of possibilities to obtain the same results by re-using OPEDAS, rather than unnecessarily creating new privacy-sensitive data that put significant responsibilities and financial challenges on your group.

DO

- Check with all investigators as to whether temporary or structural protection of the data is needed.

- Estimate the period over which the data needs to be protected and check the availability of infrastructure, funds, and procedures to enable that.

- Estimate the costs of long-term protection of the data after the experiments have been concluded and the results published.

DON'T

- Assume that data (even if there are no privacy and sensitivity issues) can just be put on 'any laptop' in the institution. You might still be violating internal procedures.

Resources: `http://dmp.fairdata.solutions/resources/zxp`

2.16.2.2 Is your institutes' security sufficient for storage?

What's up?

If data (for instance, on patients) need to be structurally protected, you need very specific infrastructure and security measures. If your institution would not have or support these (in house or in a private cloud-type environment), you should seriously reconsider whether it is responsible to collect privacy-sensitive data. Storing privacy-sensitive data is a 'profession', not something you do on the side.

DO

- Check with your ICT staff regarding all requirements to store privacy-sensitive data locally.

- Involve ethical and legal experts in the discussion in case of any doubt.

- If you intend to collect large datasets, also consider the costs and logistics related to size.

DON'T

- Assume that internal procedures developed by yourself will be sufficient.

- Assume that large datasets will be manageable by the current infrastructure of your institution. Expanding 24/7 reliable infrastructure with high security is a very costly issue and carries many specialised challenges.

Resources: `http://dmp.fairdata.solutions/resources/tgd`

CHAPTER 3

Data Cycle Step 2: Data Design and Planning

In this chapter, considerations as to the actual design of data formats, basic software, and hardware choices to be involved in the planned experiments are covered. The design of the correct data formats, ontologies, and software tools needed for your experiments or collections are a crucial step in correct scientific design and good research practice. The planning of backups, integrity, security, versioning and high performance analytics versus 'off line archiving' and many other issues that pertain to solid capacity and stewardship planning are covered here. Generating data without properly working out what will be needed to make them reusable by your own group and by others is considered a critical missing step for data stewardship in open science.

In this second step in the data cycle, data stewards play an important role, not necessarily by influencing the sort of data captured, but much more in terms of the guidance for the group as to the formats and detail in which data, metadata, and provenance should be captured. Data stewards should seek to make data optimally usable and reusable, not only in the study for which they were generated or collected, but also in future studies, not yet imagined at the time of creation of the data.

3.1   ARE YOU USING DATA TYPES USED BY OTHERS, TOO?

What's up?

Unless you do entirely novel types of research, there are likely to be multiple data formats in which such data can be represented, and formatted. Some of these may be 'exotic' and not used very much by the majority of the community, which frequently means that they will be difficult to find, map, inter-operate, and reuse. In addition, it is less likely that standard workflows will process these data formats. Especially if you intend to use the data generated in combinatorial or integrated experiments with OPEDAS, the formatting of your data is extremely important. In many cases, data in proprietary or exotic formats can be munged and recreated into more commonly used formats, but these processes are very cumbersome and error-prone. It is therefore of the utmost importance to consult the expert community and get the data in the most optimal formats for further analysis, and ultimately for reuse by your own group and others.

DO

- Always use community-compliant, supported, and sustainable data formats whenever possible.

- Turn to experts who can tell you the best formats to use for the particular data types you will create.

- Ensure you are prepared to answer questions on the use of the data (for instance, which workflows will they be subjected to).

- Choose the formats with the richest expression possibility. It is easier to leave things blank than to extend a poor data format later.

DON'T

- Assume that your data is so unique that it needs an entirely new format.

- Think that a spreadsheet with free text labels or your locally developed database is the best way to store and reuse your data.

- Format and store data in any format without keeping rich and relevant metadata and provenance.

- Throw away the original data unless you are absolutely sure that storing them has no further added value, for example, for review of experimental and analytical procedures. Not having certain pre-formatted data available may actually preclude the publication, reuse, and citation of your (original) data by others, and might also jeopardise the publication of accompanying articles.

Resources: `http://dmp.fairdata.solutions/resources/njy`

3.1.1   What format(s) will you use for the data?

What's up?

Once you know the general 'type' of data you will generate, it is important to check carefully as to the best format for capturing, processing, and formatting the data. In many cases, the capture format will be initially dictated by the software packages coupled to the instrument through which the data are measured or collected (from questionnaires to high-throughput machines). However, once the raw data have been collected, you may have many choices for how to pre-process it, and as to which format(s) to choose for the final version of the data, to go into further analysis. Considerations for choice must be mainly based on FAIR principles and community adoption of the format, which is in turn related to the availability of mapping tools, integration tools and analysis workflows in general. Having your data in easy, local custom formats like Excel with free text in cells may put severe pressure on your resources later, as significant effort may have to go into formatting with all the negative effects described earlier.

DO

- Carefully study the data formats available for the type of data you will generate.

- Study their community adoption rate.

- Check as to how much the resulting data will answer to the FAIR principles.

- Make sure the data format itself and any changes/additions you may have introduced to it are well documented and part of the (FAIR) metadata.

- Inform the provider of the data format and templates on any prerequisites in the templates or formats that would render the data non-machine actionable.

DON'T

- Use free text in any part of the data format without underlying PIDs.

- Follow instructions in data templates when you can predict this will render data not FAIR (contact the the template owner to correct).

- Ever design a data template or format unless you are absolutely sure there is no community-adopted alternative.

Resources: `http://dmp.fairdata.solutions/resources/hea`

3.2 WILL YOU BE USING NEW TYPES OF DATA?

What's up?

It might be the case that the type of data you will create is entirely new. In that case, obviously, standard formats do not come automatically. However, even for entirely new types of data, any of the existing data formats or templates may be well suited or the combination of existing formats, vocabularies, etc., will suffice. Whenever this is an option, the reuse of existing formats is preferable to the creation of entirely new data formats. There are communities specializing in these data-related topics, and it is highly advisable to work with such communities to solve your data-modelling issues.

DO

- Study existing formats and vocabularies (especially those from other disciplines as well) for suitability to your data type.

- Consider combinations of existing elements if no standard best practice exists.

- Consult experts in similar data types to make sure you did not miss anything.

DON'T

- Ever create a new data format, under the assumption that your data is unique, without a thorough study of what already exists.

- Create data types without consulting experts in the the field of related data types.

Resources: `http://dmp.fairdata.solutions/resources/ikk`

3.2.1   Are there suitable terminology systems?

What's up?

A key issue in science is to define any concept you refer to with the utmost possible precision. So, in data stewardship and FAIR data, a critical issue is the use of controlled vocabularies, thesauri, and ontologies, here generally referred to as 'terminology systems', where the concepts you refer to are defined as precisely as possible. Before embarking on the mapping of the concepts in your data to any particular terminology system, you need to study the basics of these systems, to at least know the differences between lists of controlled terms and synonyms and the actual use of ontologies where functional relationships between concepts are fixed. Ontologies assume a given abstraction of, and a particular view on, reality, one that may not always reflect the needs you have.

Controlled vocabularies and thesauri usually restrict themselves to providing terms with their defined meaning, along with synonyms (and in some cases homonyms, multiple meanings for the same symbol).

For deeper knowledge about terminology systems, and specifically about the semiotic (Ogden triangle) and how this relates to data publishing, we refer to [Mons and Velterop, 2009].

In brief: FAIR guiding principles assume that none of the concepts in your data is referred to with anything else than a machine-resolvable, persistent identifier.

DO

- Make sure you do not qualify yourself as a data expert without having a basic understanding of terminology systems, their strengths and weaknesses, and their limitations.

- Make sure that all concepts you want to refer to are covered by an existing terminology system with the highest possible community adoption.

- Make a list of concepts present in your data that are not covered by the first-choice terminology system.

- Consider using persistent identifiers from other terminology systems, following the same criteria.

- Only if concepts are not properly identified in any existing terminology system and you do need to refer to them in your data, create a new concept identity and define its meaning.

- Report the creation of a new concept + defined meaning to the nearest 'authority' you know, and preferably to the stewards of the terminology system that would be the most likely home for the newly defined concept.

DON'T

- Mix the terms 'ontology', 'controlled vocabulary', and 'thesaurus' indiscriminately in conversations and text.

- Use ontologies to refer to if you can just as well refer to a controlled vocabulary. Ontologies assume many links of the concept you refer to with other concepts, which may or may not be correct in your case.

- Create a new symbol or identifier for a concept in your data without an exhaustive search for existing terminology systems.

- Create a concept without a first attempt to define what 'unit of thought' you exactly mean by the new identifier.

- Create your own sub-terminology system without sharing that effort with the community.

Resources: `http://dmp.fairdata.solutions/resources/rzn`

3.2.2 Do you need to develop new terminology systems?

What's up?

If, after a thorough search you have to conclude that no suitable, or easily expandable, terminology system exists, you may (as a last resort only!) decide to embark on the development of a new terminology system. If that is unavoidable, you should at least follow the basic rules for the design of proper terminology systems. First of all, consider contacting stewards of existing and community-adopted terminology systems to investigate whether they are open to extending their system with the new concepts you need to refer to in your data. This will greatly enhance exposure of the new identifiers and terms you add to the community, and thus enhance adoption of your extension.

If all that does not work, start by making a very simple, one-dimensional list of concept terms you need to add to the new (addendum or) terminology system. Then, define the concepts as concisely and unambiguously as possible, adding as many synonymous symbols of the term as you can think of, also in other languages, if this is within the scope of the research. This is, for instance, extremely important in sample-banking research. Choose a proper persistent identifier approach to attach a persistent, machine-resolvable identifier to each concept-denoting term. Finally, make sure the definition can be located (so you need a URI, which is not necessarily the same as a PID). If at all possible, do all this in a community of experts in the same domain.

DO

- Make sure you have exhausted all possibilities to work with existing terminology systems or consortia before you ever start a new one.

- Try to align with other domain experts who may have the same problem, and try to make the broadest possible group (others are likely to make their own little term system again). There are fora on which to post these issues, and a global community working on community development of standards and best practices.

- Keep it simple and open, just enough, just in time, and widely announce you are doing this to try and avoid duplicate, time consuming, and cumbersome projects.

- If you need to make an ontology (add relationships), make sure your relationships (mostly 'predicate type' in linked data terms) are also well-defined concepts, so that any 'triple' of the nature [subject] [predicate] [object] is a triple of three well-defined concepts, each with a proper, persistent identifier.

DON'T

- Ever start an 'ontology' from scratch. First, make a simple list of concepts you need to refer to, and do not bother yet with their interrelationships.

- Ever make a thesaurus with hierarchical (in narrower and broader terms) relationships unless you really need that.

- Add relationships between the concepts in your list without defining these as precisely as the two concepts you want to link with that 'predicate' (see 'DO' above).

- Assume that you can capture full reality in any practical ontology. Other people will always contest the types of relationships you chose because they have another perspective.

Resources: `http://dmp.fairdata.solutions/resources/ske`

3.2.3 How will you describe your data format?

What's up?

For machines (even more than for people), it is imminently important that the format and the structure of the data you put up for reuse is well defined. In fact, once the data type, the data format, and the individual data elements are all 'understandable' for the 'visiting compute', machines will be able to reuse your data, and, if needed, transform them into other desired formats with minimal risk for errors.

It is therefore very important to include information about these elements (format, structure, terminology systems used) in your 'intrinsic metadata'. Perfectly formatted data with no information on format, structure, and terminology system may still be processed by machines (parsing all elements and 'reconstructing' what each element is all about), but it is not considered good practice to leave machines with much less information than they need, and let them figure out what the data might be about. There are several projects focusing on data provenance, extended annotation, retrieval, and citation, and you should study their recommendations (and work with them and/or their services where appropriate) to make sure that your metadata themselves have a well-defined format and contain rich information on the format of the data themselves.

DO

- Study established metadata formats.

- Keep your metadata schema and data formats as simple as possible.

- Use existing libraries and catalogues of data types and formats wherever possible.

- Record and provide the richest possible metadata format, as it is not always easy to anticipate reuse later, but minimally (in the context of this topic) include information on data format, structure, size, and terminology systems used to refer to concepts in the (meta-)data.

DON'T

- Use data formats that are not generally accepted by the community (and hence by established workflows) unless unavoidable.

- Assume that your data are structured and 'FAIR' enough in and of themselves to get away with minimal metadata. FAIR Metadata will increasingly be used by search engines to find and recommend your data, so reuse will be largely dependent on FAIR metadata.

Resources: `http://dmp.fairdata.solutions/resources/jwg`

3.3   HOW WILL YOU BE STORING METADATA?

What's up?

Always consider the use of your data beyond the original purpose. One of the issues with searching for other people's data is that they cannot be 'intrinsically indexed' as easily as, for instance, documents on the Web. For that reason, many datasets and databases go unnoticed, and are at present heavily underutilised. This does not only hold for small and highly specialised databases. Even core resources such as the Human Protein Atlas are unknown to many researchers, and much of the information in those databases is not readily available for indexing. All kinds of synonym problems preclude effective finding of such data and information sources. Many attempts are under way to develop data search engines, which are all challenged by the enormous variety in metadata standards, formats, and the data formats themselves. If all metadata were in FAIR format, the development of effective and highly performing search engines for all datasets available in the 'Internet of Data' would be infinitely more easy. Therefore, it is important that you format your metadata according to the FAIR principles.

DO

- Map all concepts you refer to in your metadata (including instruments, organisations, and contributors, etc.) to community compliant terminology systems.

- Choose a FAIR-guided format for the metadata.

- Store and 'expose' the metadata in open access environments where search engines can easily find them, even of the data they describe are not (yet) FAIR, or even highly restricted in access. The 'fact' that a dataset with specific characteristics and contents is 'out there' is a first step toward effective reuse of your data or information source.

DON'T

- Make non-FAIR metadata and assume that search engines will figure out where and what your data is anyway.

- Make FAIR metadata (in terms of format) but store them in a place where search engines have great difficulty finding them (off line, behind firewalls etc.).

- Make metadata so minimal that even finding them gives the user (frequently a machine) no real clue as to what data or information is really 'there'.

- Assume that with some 'hints' the rest will be obvious; rather assume that the most 'ignorant' machine or person will try and figure out what your dataset contains and is about.

Resources: `http://dmp.fairdata.solutions/resources/rhm`

3.3.1 Did you consider how to monitor data integrity?

What's up?

Both metadata and data themselves may be subject to change, intended or unintended. Electronic means of storing data and information does not guarantee its integrity. Apart from damage of the storage environment (to be addressed by redundancy and backups), the data itself may get unintentionally corrupted or changed. Also, data may be in formats that at some point in the future can no longer be easily 'read' by current software. In addition to those unintended changes, not all data and information is 'static' in nature. For instance, annotations of samples may have to change when the samples change. Having been taken out of frozen storage, for instance, may change characteristics, and this has to be recorded. Also, user-defined metadata of datasets may change over time when usage of the data reveals new insights about the data. In addition, there are curated and summarising information resources that intrinsically change as they are updated when new data becomes available. Therefore, data integrity, versions of (meta-)data, and updates need to be carefully monitored and regularly checked and monitored. *nota bene*: Good stewards also record when their stocks are 'outdated', corrupted, or otherwise no longer usable, and 'intelligently dispose' of useless stocks, with proper reference and communication to users of the fact that a given dataset is no longer available or has been archived.

DO

- Make sure the metadata of your dataset is kept in different places (backup) and represented on catalogues where appropriate, but keep track o- all versions the metadata.

- Keep detailed logs and provenance of whatever happened to (meta-)data and/or samples after they were captured, preprocessed, and archived initially. Especially when multiple copies of the data exist, make sure that derivative copies are not changed without recording the change in the metadata, so that people as well as machines will be made aware that they are now using a potentially changed version of the original data.

- Record all actions that happened to archived physical samples, so that future users of (metadata) and the samples themselves will know exactly which version they are using.

- Choose a FAIR-guided format for the metadata, but also make strict versioning part of the infrastructure in which you store and provide the data.

- Make (multiple) physically separated copies of the data, but keep very strict information of where data copies reside and how their integrity is guaranteed.

- Keep master lists and metadata as closely associated with the original data or samples as possible.

- Give specific people in the group access control rights and the responsibility to monitor data integrity at regular and frequent intervals.

DON'T

- Ever assume that data (let alone physical samples they may describe) will remain constant over longer periods of time without good care.

- Leave data for long periods without regular attention, only to find out that they have been corrupted or used for the wrong purposes by others.

- Let data be reused without a proper license which also safeguards data integrity rules. When workflows visit the data rather than entire copies of the set being downloaded for reuse elsewhere, workflows may make intentional or unintentional changes to the original data.

- Throw away data (even if they are corrupted or obsolete) without keeping the original 'unique identifier' to the data in the international record. People and machines should know 'the data there', what they were, and, as well as when and how they were 'taken off line.'

Resources: `http://dmp.fairdata.solutions/resources/spg`

3.3.2 Will you store licenses with the data?

What's up?

Always consider the use of your data beyond the original purpose. One of the issues with reusing other people's data is that they cannot be assumed to be reusable from an ethical or legal standpoint without explicit permission. Assuming that unlicensed data are 'free to use for whatever purpose' is intrinsically wrong, and in the case of, for instance, the pharmaceutical industry can lead to court cases later on. Therefore, whenever you publish a dataset or any other kind of information or digital object, it is important to define a license for reuse. For software many licences exist, and for data, increasingly, standard licenses are available or under development. Please note that a given license is also a defined concept, and therefore deserves a persistent identifier and a URI pointing to where the license can be studied machine-readable licenses are also under development in some areas). This means that in the metadata, the license under which the data or the workflow can be reused is 'just another PID in the right place'. Users can then specify in their search or workflow container that 'only data with the following licenses should be included'.

For instance, if you include some data in your analysis that cannot be used for commercial purposes, that decision may render your entire results not usable for commercial purposes (at least in the view of some lawyers). This means that not licensing your data at all, even if you don't care who uses them and for what purpose, is very counterproductive, and will severely undermine the actual reuse of your data by others, and in particular, by industry. It will also lower the attribution rate (usually part of the license conditions) and thus the citation and the impact score of your data.

DO

- Always carefully choose a license to be attached to your data upon publication.

- Include and clearly mark the licences PID as a concept + attributes in the metadata.

- Store and 'expose' the license as part of the metadata in open access environments where search engines can easily find the license, even if the data they describe are not (yet) FAIR or even highly restricted in access. The 'fact' that a dataset with a specific license is 'out there' is a first step toward effective reuse of your data or information source.

- Make sure, especially when you restrict use of your data, that you are able to enforce the license you choose. Licenses that are not enforceable make no sense. (Please note that the enforcement is usually not done by an individual research group but at the institutional or repository level).

DON'T

- Ever publish data without a license attached, or choose a license lightly, without considerations of anticipated reuse of your data.

- Choose a license that is not transitive (i.e., cannot be transferred with subsets of the data), but make sure its transitivity does not unduly restrict the reuse of your data.

- Choose an unnecessarily complicated license with many clauses, and wherever possible, one that is already widely adopted in the research community for either software or data.

- Restrict the reuse of data any more than absolutely necessary: For data generated with public funding, the default is usually completely open, and only restricted in any way if there are very good ethical or strategic/commercial reasons.

- Opt out of open data lightly: Most public funding agencies will request open data publishing as a default as part of their funding conditions. Usually there is an 'opt out' option, but DO NOT use that unless it is unavoidable. With the fierce competition for research grants, any element in your grant that can make it less attractive to reviewers (and keeping your data out of the public domain is certainly one of them) may cause rejection, even if it scores in the eligible ranks based on the science case. So the advice is to only restrict the reuse of data (especially for projects funded from public sources) if there is 'no reasonable alternative', and make sure you make a very strong point in your data stewardship plan about the underlying reasons.

Resources: `http://dmp.fairdata.solutions/resources/atw`

3.4 METHOD STEWARDSHIP

In the era of machine-assisted data analysis, and increasingly 'autonomous' machines and workflows addressing data, the workflows and the methods used to produce, capture, process, annotate, curate and integrate or link data are as important as the data themselves. Therefore, first of all, whatever is said about data in this book almost always applies entirely to methods and scientific workflows as well. The way in which things were done, either by people or by machines, should be carefully and comprehensively recorded. This holds for all parts of the data cycle, as much of the non-reproducibility of scientific results is due to early phases in the data life cycle, and even to inadequate descriptions of the original methods and protocols used to generate the data. This already starts with the description of the biological reagents and reference materials used in the experiment (36%), and the description of the study design (28%) before the actual 'output' data aspects start to play a role (25%), while a remaining estimated 11% in [21] is attributed to inadequate laboratory protocols. Although it might be argued that a data steward cannot be held responsible for errors introduced during the planning and the conduct of the experiment or the study itself, we can in fact consider all 'information' about every aspect of the scientific process as a form of data. This certainly holds true for the actual (machine-readable where possible) systematic description of the reagents (with unique identifiers rather than 'names', their batch number, etc.), as well as the protocols used in surveys, experiments, and other research workflow elements. So, in fact, everything that is 'said' and recorded about the entire research data cycle should be treated as 'data' and, based on the broad definition of 'data' we use in this book, algorithmic workflows are a form of 'executable data'. In other words, a good data steward will be involved in all these phases of the experimental cycle, and will have to point out the proper recording of all information that will influence reproducibility of the experiment and the quality of the data, and therefore its potential for reuse. Therefore, the questions in this section are not necessarily issues to be *solved* by the data steward, but will in many cases be pertinent to be asked of the experimental research team. One key element is that, while researchers may be increasingly convinced to refer to 'concepts in their data' with unambiguous PIDs, they may be less convinced to refer to any concept (reagents, questionnaires, instruments, software, workflow versions, etc.) with the same rigour. This is most likely causing the vast majority of the current non-reproducibility problem in scientific literature-reported studies. So, you are also the steward of (the information about) the actual research and experimental process, from 'design' to 'demolition'.

3.4.1   Is all software for steps in your workflow properly maintained?

What's up?

A frequent source of errors and lack of reproducibility is that variance in software used for pre-processing or the analysis of data. Referring back to the term 'professorware' in the introduction, many software packages and algorithms are not even reaching that level, and are custom-made programs in coding languages that the informatician of the department happened to master.

The academic culture to press data scientists and engineers to publish about new algorithms and software packages drives the bias towards multiple, custom-made workflows for identical research steps.

However, to be the basis of reproducible results based on the same data, the workflow and all elements it uses (such as, for instance, thesauri for text-mining, lists of reference data, etc.) need to be exactly the same. However, in many cases, one or more of the components are not well documented, maintained, or versioned (people you do not control decide on updates without even informing you). So, consider writing new software or algorithms for data analysis without the need to, as a first capital sin for good data stewardship. Unless you are the very first to run a particular type of experiment, more likely than not, software and workflows/pipelines for the type of processing and analysis you need will already exist.

Even if you use 'existing' open source (OS) software produced in your scientific community, first verify that there is proper support, versioning, and documentation about the code. If not, you run a serious risk of running into irreparable reproducibility problems very soon. In many cases, workflow decay in the public sector is a very serious problem. Next to that, many OS workflows are based on serial running of Web-service-type components. If one of those is 'off-line' or changed (without proper management and notice), your workflow will either not run properly or give unexplainable variations in results. It is therefore of the utmost importance and core business for data stewards to ensure that the software components you decide to use are of sufficient stability and quality to take the risk of subjecting your valuable data to them. Even for commercially provided and professionally engineered software and pipelines, there is always the risk that the company supporting the tool will have suboptimal versioning and support agreements in place and/or disappear from the scene. It is well established that tools which have already been evaluated as improper for the task they are used for continue to be used (and pass peer review) for years afterwards.

DO

- Make sure you have studied the landscape of tools and services that meet your quality standards in terms of performance, stability, versioning, documentation, and sustainability. In several fields, there are tool registries and comments on the issues with particular tools or workflows.

- Choose only the most commonly used and validated tools available, and even in that case, document very carefully for yourself the date, the version, and the conditions under which you have used the tools in your data processing activities.

- Also check 'underlying components and plug ins' for their stability, etc. As said before, if a text-mining module changes the version of a terminology system between two runs, you will find very different results even if the text-mining software itself is perfectly stable.

- Contact the 'owner' of the software or tool whenever there is even the slightest doubt about all aspects mentioned above, and make sure that future use of exactly the same composition and version can be guaranteed. When workflows or tools were only published recently, and there is no evidence that they will be properly open sourced or otherwise sustainably provided in the future, think twice before using them.

DON'T

- Treat tools with any less care then the data themselves, which means that metadata about the tool, how you used it, which version, which components, when and with what exact input and output data should be captured and stewarded with the same care as described for data.

- Develop any software or other data-tooling unless there is really no alternative. For instance, building a new tool that is 20% faster than a commercial alternative, just for the fun of it (or for a publication) and using that custom tool on valuable data other than for tool evaluation *per se*, should be considered a no-go area for data stewards.

- Think that OS tools (and your potential additions to them) are well supported just because they are OS. Unless there is a foundation or a company with some sustainability that supports the tool (OS or not), it is very likely that some time later, when the 'PhD student who made it moved on', the tool will either just go off line or give crappy or very different results.

Resources: `http://dmp.fairdata.solutions/resources/brz`

3.5   STORAGE (HOW WILL YOU STORE YOUR DATA?)

Data storage is no longer trivial. Especially when data is larger than 10 TB, the usual storage facilities in research institutes are not ready to handle that size of data efficiently, particularly with guaranteed up-time and download facilities. It should be noted here that purely storing and archiving data is quite a different thing from putting them up for 'reuse' (access, download). The latter is an 'order of magnitude' more complex and labour/infra-intensive, and therefore also very significantly more expensive. Storage capacity and condition planning is a serious activity in good data stewardship, and goes way beyond 'enough disc space'. It includes questions to be asked about backups, required up-time, speed of 'recovery from archive' into actionable state, usage requirements, off line and online availability, etc. The very basic requirement is obviously safe, reliable, and redundant (distributed) infrastructure, but that is only the start.

3.5.1   Storage capacity planning.

What's up?

Storage issues do not necessarily scale linearly with the size of the data in 'bytes'. The complexity and the nature of the data should be taken into consideration, and at certain sizes there might be a sudden 'breaking point'; for instance, your internally available infrastructure cannot handle it any more, or there is an institutional policy concerning maximum data sizes.

DO

- Check the availability of sufficient and reliable storage capacity in your institute, and discuss with the department responsible for it.

- Make an upfront calculation of the full costs of the initial storage, backup, and long-term preservation of the data with the experts in the group.

- Inform the local ICT experts about the intended use and reuse of the data over time, and discuss the consequences of that plan for the sort of storage and availability needed. This will largely determine the costs.

- Consider all options to reduce data sizes without information loss and smart sharing options.

DON'T

- Assume that the storage in your institute, even if it is very large, is 'just there'. Storage is no longer something that simply will become cheaper so fast that it can be considered marginal cost.

- Surprise your institutional ICT staff with large datasets they are 'supposed to store', as it is their task.

- Consider storage as purely 'archival', but always consider the use and reuse requirements at the same time, as they may significantly influence the choices made (for instance, can data be stored on tape or not?)

Resources: `http://dmp.fairdata.solutions/resources/yqy`

3.5.1.1 Will you be archiving data for long-term preservation?

What's up?

Long-term archiving and preservation is clearly distinct from 'short-term (re-)use of the data for the experimental procedures immediately following its generation. Even for short-term use, data need to be stored and backed-up in different places, with the appropriate safety and access considerations.

However, when it is decided to keep the (reusable) data for prolonged time periods, the scale of these issues increases. The conditions under which others can use the data later on may differ (legally or practically) from the way in which the data was used for its original purpose. Therefore, access, constant availability (or not), and many other issues may have to be considered. It is good practice to anticipate the issues associated with the long-term preservation and reuse up front, and budget for them as part and parcel of the data-steward plan when the data-generating experiments are designed and planned.

DO

- Consider data 'long-term re-usable unless. . . ' It is easy to assume that data (small or large) are only of interest to the person or the small group that generates them. However, even if you think (correctly) that the data will be never useful for any other experiments, the very minimum requirement is that they remain available for others (including reviewers requesting re-running of experimental or analytical workflows).

- Always keep the persistent identifier of the dataset available. Even if the data are deemed to be of no use any longer, and they are taken off line or even destroyed, the fact that the dataset was there (and may have been cited by others) needs to be traceable.

- Continue to update the metadata of the dataset to ensure that its fate (including reuse) over time is always traceable. It is highly recommended that you create an 'explanatory file' describing what the data was and why it was taken off line when the decision is made to no longer preserve the data.

DON'T

- Assume that long-term preservation is something to 'worry about later', and upset the ICT colleagues with *ad hoc* and urgent solutions when it appears the data are 'valuable after all'.

- Store dataset (small or large) without obtaining a registered PID (for instance, a DOI) for the dataset.

- Wait with generating a PID until the data are preserved, but make it a routine to do this at data generation time.

- Let data change over time (even just location) without making sure it remains findable and accessible.

Resources: `http://dmp.fairdata.solutions/resources/kjp`

3.5.1.2 Can the original data be regenerated?

What's up?

In some cases it might be cheaper (and acceptable) to regenerate data rather than storing them. Two examples: It may soon become cheaper to 're-sequence' a genome than to store it for 10 years. Also, text-mining the same corpus of text with the same tagger and the same thesaurus, should in principle give the exact same result when repeated at any time. However, in both examples, a number of assumptions would have to be made before a decision would be made to re-generate the data rather than storing the first version for an extended period of time. First of all, the technology should not change; sequencers get more reliable by the day, and therefore, may give different results, and the 'old sequencer' may not be in your possession any more by the time you want to generate the results. Workflows are not necessarily stable, but more importantly even 'stable' substrates (a genome of a living individual or a corpus of text) may not be as stable as you think. Changes to a text corpus may occur unbeknownst to you, but also, the somatic mutation rate in the genome of a living organism is not insignificant, and therefore, a new sample of cells from which to take DNA may give different results. Even if the DNA sample was stored in a 'preserved state', there is no absolute guarantee that later re-sequencing of it will give exactly the same result. So, in all cases, the decision to 'regenerate versus store' is a deep-scientific, method discussion in the group, and not a 'trivial decision'.

DO

- Consider all angles of the problem, including deep domain knowledge issues like the ones exemplified above when a decision for regeneration of 'identical' data is to be considered.

- Only consider this option if the long-term preservation is problematic due to size/costs or other aspects.

- Always keep careful records of whether data are indeed 'exactly the same' (as far as that can ever be guaranteed) or 'supposedly the same, but regenerated from the same substrate with the same methods.

DON'T

- Lightly assume that data can be easily regenerated, even if that seems to be apparent with only superficial knowledge of the scientific subject.

- Re-generate data and archive them under the same PID as the original dataset, assuming that there will be no differences.

Resources: `http://dmp.fairdata.solutions/resources/ixr`

3.5.1.3 If your data changes over time, how frequently do you do backups?

What's up?

Data may change over time. Not only curated databases that get updated 'as a routine', but also sources where errors may be spotted and corrected. This is not necessarily only pertaining to wrong values, but maybe also correcting misplaced values in the wrong column, etc. It is extremely important to record any post-capture and initial archiving changes and to also reflect these corrections in an immediate new version and backup of the changed data. The provenance should also enable to 'go back' to the original data, even if these were deemed inappropriate in hindsight. Reviewers of the conclusions should, for instance, be able to trace why initial conclusions were revised.

DO

- Backup old and new versions of data.

- Keep exact and rich provenance of all changes post-initial generation.

- Explain why data have been adapted, curated, cleaned, etc. and also make sure workflows are properly instructed as to which version of the data to use.

- Give each new version of a dataset a new PID.

DON'T

- Ever change anything in recorded data without informing the original data owner.

- Make changes (even obvious and minimal corrections) without recording and documenting that change.

- Make a change without preserving both the old and the new versions of the data with the appropriate metadata and backups.

Resources: `http://dmp.fairdata.solutions/resources/tgk`

Are you using backups for restoring files that were accidentally deleted or changed?

What's up?

Even if changes (including unintended loss) of parts of data spark 'restoration' from backup files, especially if those were not under your own control and you cannot guarantee that the backup was still fully identical to the original, you need to record and document that procedure. If people will reuse the restored data under the assumption they were using the original, and they find unexplainable differences in their results, they need to be able to trace this back to the (potential) errors that were introduced during the restoration process from external backups.

DO

- Record and document everything that happened to the original as well as the backups.

- Require the internal and external parties that take care of your backups to inform you immediately if there is any reason to assume that the backup does not resemble the original 100% any more.

- This could include backups of identical subsets of, for example, toxicity and tissue samples (accidentally thawed?).

DON'T

- Think that backup only starts when data is archived: You may lose them during experimentation.

- Assume that backup files of archived, reformatted, or even destroyed data and their metadata deserve any less attention and care than those of active data.

Resources: `http://dmp.fairdata.solutions/resources/eky`

3.5.2 When is the data archived?

What's up?

The term 'archived' is usually reserved for the process when data 'retires' from the project for which is was originally generated. However, in the data-driven science era, such data may be 'called from retirement' at any point in time. Still, we keep the term 'archiving' here for the process that follows after the intensive use period in the data-generating project. Archiving for preservation only (let's call it 'taping') may render the data integer recoverable but not necessarily immediately reusable.

We should still consider these data part of the FAIR ecosystem in case the FAIR metadata are still exposed, and both people and machines are able to 'find them', and find out what their accessibility level is (needs to be recovered from tape, needs personal contact, etc.) even if it will take considerable effort to reconstruct the data in readily interoperable and reusable format. It should be decided as early as possible in the process how and when the data will be archived, in what format, on what carrier and how they will be protected against calamities, unwanted or improper use, and theft.

DO

- Decide with the team what the best time point and method are to archive the data for long-term preservation.

- Keep rich documentation of the procedures followed to format the data for long-term archiving, and consider potential sources of error and change introduced by these procedures.

- Record the exact time of archiving and the authority (repository?) that takes over responsibility.

- Give that archived version of the data a new PID (and if you destroy the 'working version' of the data, keep that PID for later reference).

DON'T

- Mix archiving with keeping data in store for reuse in the same study cycle or for further processing.

- Mix up archiving with keeping data in 'active state'.

- Consider data that are 'somewhere on a disc' as being properly archived.

Resources: `http://dmp.fairdata.solutions/resources/rht`

3.5.3 Re-use considerations: Will the archive need to be online?

What's up?

There is a very serious difference between 'off line' archives (obviously with online FAIR metadata) and online (and ready-for-reuse) data archives. Actually, it is considered better to reserve the term 'archive' for the situation where the actual data are 'off line', and if found, based on the FAIR metadata can be retrieved from the archive and made 'reusable on demand', and to distinguish those 'archives' from 'high-performance reusability' (HPR) environments where data is kept in a 'poised for frequent reuse' format. The latter puts many more requirements on the bandwidth, up-time, and support of the infrastructure in which the data are offered.

DO

- Always make the distinction between off line archiving for reuse upon request and a 'high performance reusability' (HPR) environment.

- Realise that the latter may be an order of magnitude more costly than the former.

- Budget adequately for either form of 'archiving' after the conclusion of the experimental procedures and analytics with the data that were originally generated and interpreted.

- Consider the potential need of users to have 24/7 access to the data, which is usually beyond the capacity and mandate of academic institutions.

DON'T

- Put data in long-term HPR environments unless you have strong expectations or evidence that they will be reused intensively.

- Assume that you can offer data for reuse by just putting them somewhere for people to download. The offer of reusable data comes with stewardship responsibilities and sometimes with many questions for which you have to be prepared.

- Consider your group a 'local HPR' node without clear mandates and personnel. You may want to budget for 'handing over' your data to a trusted and professional repository or HPR environment.

Resources: `http://dmp.fairdata.solutions/resources/ybd`

3.5.4   Will workflows need to be run locally on the stored data?

What's up?

Many datasets will be too large to be effectively and economically 'shipped around' even if they are offered for reuse. In addition, there may be privacy, legal, or commercial restrictions that prevent the data from physically leaving your internal repository. These issues are extremely important and should be carefully studied at the early stage of data generation, as they may influence your generation, licensing and stewardship choices. If data cannot be 'shipped', the environment in which you offer them for reuse is critically important. A 'download server' where people can just 'take the data for their purpose' is very different from a place where 'workflows' (compute elements) can come in and do their calculations and analytics locally on your dataset. Security and authentication issues (here pertaining to visiting workflows) are very different, and also, there should be sufficient compute power directly associated with the data in order for the workflows to do their work efficiently. The licensing to the data and what may or may not be done with the results of the local computation have to be very clearly defined, and a support mechanism for the *'in situ'* reuse of the data should be in place. Again, this may be outsourced to a trusted party. Also consider that the metadata of your dataset may be provided to external parties in open access, as opposed to the data itself.

DO

- Discuss the licensing and security issues related to the data to be produced before capture wherever possible, as these constraints may influence your metadata capture strategies, your formatting and the budget of your experiment very significantly.

- Consult with experienced people in the area of workflows and distributed learning to understand what the requirements would be to make your data optimally reusable for 'visiting software' or workflows in general.

- Test run workflows on your data to confirm that they are indeed 'accessible' for workflows.

- Include rich metadata on the format, the constraints, and the usability of the data; for instance, which identifier scheme was used internally and warnings that the data are not 'ready for use' for particular workflows.

DON'T

- Capture data without having considered the perceived needs for reuse, and in particular, whether that is classical data sharing or access to your data *'in situ'*.

- Assume that 'someone else' in your institute will take care of the infrastructure and support the reuse of your data by third parties, but ensure that this is well organised, or budget for internal or external services.

- Waste time of the support people (internal or external) by not being prepared to answer questions about reuse conditions (download versus *in situ*, up-time guarantees needed, security level, licenses, etc.)

Resources: `http://dmp.fairdata.solutions/resources/dta`

3.5.4.1 Is there budget to enable supported reuse by others (collaboration/co-authorship)?

What's up?

It is very important when planning for the adequate budgeting of your data stewardship plan to think beyond the use of the data for your own study. Here we re-introduce the term 'data publishing'. As we should treat reusable datasets with the same care as we are used to for research articles, we need to make sure that optimal use by others, citation and sustainability, we properly budget for and otherwise cover. For research articles, it has long been accepted that external parties take part (for a fee in open access) in the formatting, redaction, peer review, publishing, and preservation of your narrative article and the accompanying resources (supplements). Think about data in the same way. They are a valuable output of your research and should be 'published' in their own right, regardless of any articles (sometimes more than one) you might want to base on the data. Obviously, the description of the data should be rich enough to make them actually reusable by others. That means aspects of findability (including good metadata and a persistent identifier), accessibility (open access, restricted, licensing) and downloadability or '*in situ* accessibility' for analytics. In the *article* sphere, interoperability of the outcome was largely related to 'readability for other human users', and this again is largely confined to proper language, rhetoric, and formatting, issues that are usually co-judged and improved in the 'peer-review' process. However, interoperability of data is more complicated. Text is a nightmare for computers, as already described, so machine-readable-and-actionable met a data formats and data formats are critical. For most data types a standard format, and, also, increasingly metadata format are available and these should be used wherever possible. Human reviewers will increasingly be unable to 'check every line' of your big datasets and therefore will have to rely on the formats and standards you used. We expect to see more and more computable 'quality check and integrity check' tools in the data space, but these are not yet available for all data types.

This all means that your study plan (or proposal) should adequately budget for data publication in FAIR format. Tools are under development to certify your (meta)data as FAIR, but these are in their infancy, and therefore, you may need to spend considerable time, effort, and funds to get your data in the correct format, to have it checked and to actually publish it. The good news is that these publishing costs are mostly 'eligible costs' in research proposals, and they also include the long-term archiving of these data. Once data are in the correct format and 'static' in size and shape, the long-term storage can usually be easily budgeted and justified.

DO

- Carefully budget for the publication of your data.

- Check various data publishing providers and check whether they are trustworthy (for instance, approved/certified by the funder of your research).

- Compare prices of different publishers, and also, specifically whether they include long-term preservation of the published data.

- Restrict this part of your budgeting effort to the actual 'archiving' as a basic preservation cost.

- Budget separately for potential storage and copies of the data in HPR environments, which might be very worthwhile in terms of increasing reuse and citations of your data, and thus its 'impact'.

- Check whether your institution has a data access committee or an equivalent body.

DON'T

- Mix one-off formatting, review, quality checks, etc., with real 'publishing' of the data in a format and environment that will make them FAIR (with the main consideration being machines, the prime 'users' of digital large datasets).

- Assume that an open source, open access academic repository is always the best option for the funder and the actual reuse of your data. Many of these are not findable or sustainable, both preconditions for FAIR.

- Mix publishing fees and budget with budgets for reuse (by others). Third parties who want to reuse your data would have to budget for reuse costs (such as download fees, re-formatting, etc.) in their grants. It is not your responsibility to take part in these costs, it is your responsibility to offer the data for reuse under well-defined conditions (which might in some cases be very restrictive or even commercial).

Resources: `http://dmp.fairdata.solutions/resources/kqh`

3.5.5   How long does the data need to be kept?

What's up?

A good steward will not keep the goods to be taken care of beyond their 'expiration date'.

For data stewards, this means that data (or subsets, such as intermediate formats, original images, etc.) do not necessarily all have to be stored and/or published. This is not a trivial issue. For some large datasets, it is far from easy or cheap to 'just keep everything'. So, the first consideration here is to determine which phases of the data generated and processed need to be 'kept forever in principle'. This means that even in the process of generation and processing of the data, there might already be files that can be disposed of responsibly. Next, some data can be zipped in the sense that they can be stored in a reduced format without information loss or the introduction of errors. An example of such a steward-decision is the question of either storing the entire sets of reads and the full imputed genome sequence of an organism as opposed to just the 'variations from the reference genome'.

In addition to these early process decisions, there may be a point when archived data appear to be not reusable anymore. It is important to distinguish here between 'not used anymore' and 'principally unusable'. If data appear to be corrupted, outdated, falsified, or in other ways misleading or useless for further research, there may be a point where the data steward decides to dispose of the data.

Even in that case, the metadata and the persistent identifier should be kept, as the community should be able to trace the original data and be aware of/refer to the fact that the data are no longer available.

DO

- Discuss a long-term data sustainability and reuse plan with the team for each dataset generated, downloaded, or acquired.

- Plan for a budget and regular 'expiration options' meetings for all datasets under your stewardship.

- Transfer these plans, resources, and responsibilities explicitly to the trusted third party to which you may give stewardship over your data.

- Make sure that even if data are 'destroyed' or 'tape-archived', the metadata as well as the unique (citable) identifier always stay FAIR.

DON'T

- Mix reusability with actual reuse: There are many examples where intrinsically valuable data had not been used for many years and suddenly appeared to be crucial for a particular study or decision. Similarly, there are examples of data that have been lost and would now be very valuable.

- Ever throw away data, metadata, code, or tools without informing the team and discussing these decisions. It is not always easy for the data steward to determine the actual expiration date of data.

- Keep data (even small sets) when there is clear evidence that they are corrupted, false, wrong, or obsolete for other reasons, even if there is no 'storage' or financial reason to delete them.

Resources: `http://dmp.fairdata.solutions/resources/kdp`

3.5.6 Will the data be understandable after a long time?

What's up?

Digital objects are not protected from decay or from lack of action-ability. Have you tried to read a floppy disk recently? The evolution of data storage, retrieval, and processing is developing so fast that, contrary to what a lay person may expect, digital objects become obsolete (unreadable) much faster than classical written text on paper or on microfiche. Communication in human language has obviously evolved as well, and reading text of many ages ago is not without its difficulties: not only because of changes in spelling and style, but also because of semantic drift.

However, recovering files only decades old that have not been updated to current formats is already a challenge in many cases. Therefore, especially for those datasets that are too large to effectively store and reuse in 'PDF' type settings, a long-term plan with regular 'checks' for readability and reusability is needed.

DO

- Discuss a long-term data sustainability and reuse plan with the team for each dataset generated, downloaded, or acquired.

- Plan for a budget and regular 'expiration options' meetings for all datasets under your stewardship.

- Transfer these plans, resources, and responsibilities explicitly to the trusted third party to which you may give stewardship over your data.

- Make sure that even if data are 'destroyed' or 'tape-archived', the metadata as well as the unique identifier always stay FAIR.

DON'T

- Mix reusability with actual reuse: There are many examples where intrinsically valuable data had not been used for many years and suddenly appeared to be critical. Similarly, there are examples of data that have been lost and would now be very valuable.

- Ever throw away data, metadata, code, or tools without informing the team and discussing these decisions. It is not always easy for the data steward to determine the actual expiration date of data.

- Keep data (even small sets) when there is clear evidence that they are corrupted, false, wrong, or obsolete for other reasons, even if there is no 'storage' or financial reason to delete them.

Resources: http://dmp.fairdata.solutions/resources/zmu

3.5.7 How frequently will you archive data?

What's up?

The decision to archive data for later reuse (by yourself or others) depends on a lot of variables. It is first of all important to understand and agree on how frequently and for what exact purpose the project partners will need to access the workspace where the data are residing. Some data may need to be mounted for immediate use all the time, or remote mounting may be needed. As a remote mount uses a network file system (NFS) to connect to directories on other machines so that they can be used as if they were all part of the user's file system, this option may only be viable for relatively small datasets. There may be many steps of intermediate data that need to be considered, or, that need to be stored for later reference, review and reproducibility checks, but these need not necessarily be all mounted all the time. Obviously, when you copy data to local work spaces, they may or may not need the same level of performance, security, and network speed as the main workspace. Also, if there is no 'local expertise' at the sub-workspace, you need to plan for support.

DO

- Discuss a long-term data sustainability and reuse plan with the team for each dataset generated, downloaded, or acquired.

- Plan for a budget and regular 'expiration options' meetings for all datasets under your stewardship.

- Transfer these plans, resources and responsibilities explicitly to the trusted third party to which you may give stewardship over your data.

- Make sure that even if data are 'destroyed' or 'tape-archived', the metadata as well as the unique identifier always stay FAIR.

DON'T

- Mix reusability with actual reuse: There are many examples where intrinsically valuable data had not been used for many years and suddenly appeared to be crucial. Similarly, there are examples of data that have been lost and would now be very valuable.

- Ever throw away data, metadata, code, or tools without informing the team and discussing these decisions. It is not always easy for the data steward to determine the actual expiration date of data.

- Keep data (even small sets) when there is clear evidence that they are corrupted, false, wrong, or obsolete for other reasons, even if there is no 'storage' or financial reason to delete them.

Resources: http://dmp.fairdata.solutions/resources/bpp

3.6   IS THERE (CRITICAL) SOFTWARE IN THE WORKSPACE?

What's up?

As argued before, the separation between data and 'software' (executable code) is close to blurred in data -driven science. In many cases, therefore, your experimental workspace will hold your *de novo* research data, OPEDAS, and software packages that you use to process and analyse the data.

This means that 'active data' like executable code may influence the workspace in a different way from 'static' data. Software packages that are open source and not properly supported may seriously disrupt the workspace and cause all kinds of trouble. Also, software used in the workspace may carry with it certain licenses that render the processed data unusable for the purposes you had in mind with them. First of all, it is therefore critically important that you never mix the loaded data for processing and analysis as they are mounted in the active workspace, counting them as 'one of the copies' or even a backup. The principle attitude must be that data that are actively used in the workspace will at some point be corrupted or lost due to unforeseen events. But also software itself (when active) may become corrupted and may need to be restored from another source in order to rerun processes and analytics.

DO

- Keep 'static' and safe backups of both data and software that is mounted in the workspace.

- Conduct regular integrity checks on the data elements in the workspace to prevent unexpected outcomes due to corruption of data, software, or processes. Unnoticed events may seriously slow down proper experimentation, analytics, and interpretation of data, and you as a steward cannot expect the rest of the research team to notice.

- Make sure that versions of the software and workflows (including plug-ins like vocabulary services) do not change unexpectedly (academics are notorious for changing things without notifying potential external users).

- Treat your workspace as a 'stand-alone' and time-delimited environment where you run stable data and processes (so, for instance, no remote mounting of thesauri of Web services you do not control).

- Keep very careful records of versions and potential issues of all elements (data as well as software) that were used to run a particular process at a particular time. Basically treat each 'run' of a workflow system as a 'batch' with a unique identifier and rich provenance.

- Make sure you can always safely and quickly restore an 'identical' workspace (although always a new 'batch' identifier will be needed) from the archive.

DON'T

- Ever regard the active workspace, its data, and its software as 'another backup' of your data.

- Produce and deliver data from the workspace back to the team without carefully monitored and described 'provenance' and a 'batch' identifier.

Resources: http://dmp.fairdata.solutions/resources/cbq

3.7 DO YOU NEED THE STORAGE CLOSE TO COMPUTER CAPACITY?

What's up?

There is an ongoing transitional debate about distributed trust and learning networks. This not only pertains to scientific realms, but is critically addressed in next-generation solutions to bit coin, block chain, and other approaches. The essence of the Internet is local transactions that are internally and intrinsically robust (including accepting error prone issues and redundancy only where redundancy makes sense). This trend will rapidly influence 'local' versus 'distributed' compute and learning. So, in any case, for a small or a large study, you have to consider the possibilities of local versus distributed computing. The options range from using smart-phones, to distributed supercomputers heating homes connected via glass fiber to exascale computer facilities. It is not a 'given' any more that massive compute and analysis jobs have to be done centrally and within firewalls even when sensitive data are involved. If we define the 'cloud' as 'other people's computers' in general, most of what you need to do may be perfectly feasible 'in the cloud'. However, there may be reasons to do things locally and maintain expensive compute infrastructure. Still, never buy locally maintained hardware if there are better and cheaper options, 'just for the sake of controlling it'. No one generates electricity or clean water any more unless via a backup generator or in cases where the water needs to be of exceptional quality that cannot be delivered from external sources. Compute and storage rapidly becomes such a 'general commodity', and using it externally or in the 'cloud' becomes a major issue to address. The core question is: How much 'compute' capacity needs to be immediately associated with the data? Recognising the speed of current connections in your network is obviously a major aspect here. This again pertains to whether you plan to send massive datasets over networks (where network speed may still be a major cost and time issue), or 'lightweight' workflow containers that visit data *in situ* to do relatively restricted compute jobs.

There are emerging trends that show that many more scientific questions can be answered by distributed analytics and learning than we would intuitively believe. Therefore, it is considered bad data stewardship if you burden our institution with internal hardware and software issues that can be done by trusted third parties much better, cheaper, and safer.

DO

- Study and consult carefully on cloud-based options for the storage and compute you will need.

- Compare different cloud services providers for pricing, service, and security (bigger is not always better).

- Make sure you are allowed to 'send data around' before going down that path.

- Consider that upload and download pricing of regular cloud providers will be considerable if you deal with really large datasets.

- Balance the costs you will be incurring using a cloud provider with the costs of local facilities.

- Calculate the risks of procurement, maintenance, support, expertise, and renewal of machinery and infrastructure in your own institution as well.

- Choose the right local/distributed approach for each individual workspace and 'batch' of analysis (not a one-size-fits-all approach).

DON'T

- Just assume that the storage and data processing will be done 'in house' without a very strong justification for that (major costs and logistics involved).

- Buy or hire much more computer space and power than you actually need.

- Consider every compute job a 'compute and data altogether' situation.

- Confuse 'large-scale compute problems' with 'large-scale' local infrastructure.

Resources: `http://dmp.fairdata.solutions/resources/wia`

3.8   COMPUTE CAPACITY PLANNING

Compute-capacity planning is no longer a straight line between 'size of data' and: *CPU's needed to analyse them properly = size & price of my computer.* There is an entire science and industry developing around smarter ways to deal with data, information, and *in silico* learning. It would be wise to study the basics of these developments and consult with top-level computer scientists or engineers in your institution or beyond, to make sure you are not lured into buying or using old-fashioned and overly expensive compute and storage facilities.

3.8.1   Determine needs in memory/CPU/IO ratios

What's up?

The ratio of storage to compute related-capacity will vary per project and data/analysis combination. However, at a very general level, the balance will only vary within certain boundaries (exceptions will always be found). Regardless of whether massive datasets are locally collected, stored, and processed by massive amounts of CPUs or thousands of distributed datasets will be visited by workflows to do relatively simple compute jobs on these data, there will always be a need to determine the size of the data, the complexity of the individual compute job, and consequently, this balance between storage and compute. If you do not properly study, pilot, test run, and decide on these issues based on current expertise and evidence, you may 'buy' stuff you do not need, and put a heavy maintenance burden on your institution/colleagues. Conversely, if you underestimate the complexity of these matters, your study may face severe hurdles and delays or even data loss, or corruption of data down the road. So, take the appropriate time and measures to make sure the decisions made are the best ones, based on current knowledge and funding.

DO

- Study and consult carefully on cloud-based options for the storage and compute you will need.

- Compare different cloud-services providers for pricing, service, and security (bigger is not always better).

- Make sure you are allowed to 'send data around' before going down that path.

- Consider that upload and download pricing of regular cloud providers will be considerable if you deal with very large datasets.

- Balance the costs you will be incurring using a cloud provider with the costs of local facilities.

- Calculate the risks of procurement, maintenance, support, expertise, and renewal of machinery and infrastructure in your own institution as well.

- Choose the right local/distributed approach for each individual workspace and 'batch' of analysis (not a one-size-fits-all approach).

DON'T

- Ever just assume that the storage and data processing will be done 'in house' without a very strong justification for that (major costs and logistics involved).

- Buy or hire much more computer space and power than you actually need.

- Consider every compute job a 'compute and data are all together' situation.

- Confuse 'large-scale compute problems' with 'large-scale' local infrastructure.

Resources: `http://dmp.fairdata.solutions/resources/ijn`

CHAPTER 4

Data Cycle Step 3: Data Capture (Equipment Phase)

In this relatively short chapter, the issues that relate to methods and instruments that will be used to capture new data, and functionally integrating them with other data are briefly addressed. The quality of data begins with proper experimental design, but is also highly dependent on how the data are captured at the source. As explained earlier, significant parts of non-reproducibility issues are directly related to insufficient details on measurements, and lacking or imprecise information on reagents, instruments, and other elements of the data-creation and capturing process. When data are published, the metadata should also be, as much as possible, readable by workflows that reuse the data. Therefore, ideally, all elements that may influence the reproducibility of results and conclusions drawn from the data in earlier use cases should be part of FAIR metadata. Metadata should also richly describe issues related to sometime proprietary software and data formats coupled with commercially available instruments, but also issues related to the recovery of data from earlier formats; for instance, social media or electronic health records. The role of the data stewards in this step of the research data cycle is probably a bit less central than in other steps, but the actual capture of the data, the richness of metadata (of which the values may have to be captured as well), and other issues that will influence the next steps in the data stewardship cycle need the continuous attention of the data steward in the team.

Once the study or experiment you are undertaking is carefully planned in terms of the use of OPEDAS and your own data to be captured, processed, stored, linked, and analysed, the actual process of data download (OPEDAS) and upload into the workspace, as well as the '*de novo*' data capture, is ready to begin. In this step of the data stewardship cycle it is important - and even more now than ever with the rest of the research team in the loop - that you consider all the issues that come up during the actual process of data creation and capture. In many laboratory settings, this step is heavily dominated by instruments: not only in typical laboratories, but also in sensor-based studies in the environment, space, etc. In many cases, these instruments measure a particular range of parameters on input of features, material, or data with which they are fed. Obviously (but a bit outside the scope of this book), the quality of measurements is heavily influenced by the settings, the quality and condition of the instrument, the reagents used, etc. So, these parameters are definitely part of the (meta)data you have to record during the process, However, as is even more deeply at the core of data stewardship, the truism 'crap in, crap out' is relevant at his stage. In other words, good data stewardship is not an 'interrupted process' while the carefully planned experiments are actually conducted. Good data stewardship starts with sitting next to the designers of the study, advising them about the data 'after they have been captured', the formats and metadata, etc. Anticipating the needs for data storage, compute and analysis, as well as long term preservation for reuse, are also obvious data stewardship issues. There may, however, be the misconception that during the actual data creation or capture and collection, the data steward would be less involved or even 'out of sight'.

In fact, it is becoming more and more clear that suboptimal methodology during sample preparation, experimental conditions, batches of reagents and laboratory-ware, contaminated cultures, wrong or mutated cell-lines, and many other elements can be the causes of non-reproducible data, and entire experiments. The instrument itself may have flaws, but in principle it will just 'measure what it is told to measure' and will produce a standardised output, frequently determined by the manufacturer, rather than by the experimentalist or the data steward *per se*. Such flaws, as well as sloppy methods in the laboratory, or during other data-generating and capture processes (down to questionnaires) are largely beyond the direct influence of the data stewards. Yet, there is also a major data stewardship task in this critical phase of the research process. Knowing that unexpected, and initially undetected anomalies in the process may heavily influence outcome, the data steward has a major 'recording' responsibility. There is a grave danger that (especially inexperienced) experimental researchers fail to properly record what they are doing. This is not a matter of proper paper or electronic notebooks, it also pertains to the decision as to what data to capture 'about the process' and, in particular, tacit issues that would easily escape the regularly recorded parameters and values.

So, deep knowledge about where the data come from, what they were meant for, and what the format will be when the measurements first leave the capture process as 'raw data' are needed. But the continuous recording of batch numbers of reagents, the materials used, and other conditions that may influence the resulting data are also crucial. These 'provenance' and process elements are key data about the experimental and data-capture process that may prove crucial, especially when unexpected outliers or unexplainable values are found in the results. Being able to 'track your steps' is a critical element of good research practice, and in particular, when your research is aimed at 'decisive' data (such as in clinical trials, or other registration, and certification-related studies), this tracking may be an absolute prerequisite. Therefore, the data stewards should feel co-responsible for taking the utmost care that if the almost unavoidable aberrations occur, and 'obvious errors' appear in data that leave an experimental instrument or process, the apparent mistakes or other causing agents (such as reagent batch) can be properly traced and corrected if needed. This is critical, as aberrations may just be unexpected findings that are indeed showing a fundamental flaw in the study setup or the hypothesis itself, and these need to be, as clearly as possible, determined and separated from detectable errors introduced by experimental misconduct or reagent/sensor-based errors. Therefore, even keeping track of the maintenance status of machines, versions of workflows used, and potentially changing conditions are part of the task of data stewards. In fact, anything that can influence the nature and the quality of the data before they enter the post-capture analytics process is part and parcel of good data stewardship, or more precisely, of good research practice. Obviously, this is again inseparable from capturing and connecting rich and powerful metadata to the process as well as to the data it generates. Here, we will not address the precise experimental elements to be taken care of, as these are highly domain specific. We will once again focus on the generic data-capture and quality aspects.

However, a good data steward should sit down with the experimental team and discuss all possible sources of error and non-reproducibility up front, before even starting the data-capture process.

4.1 WHERE DOES THE DATA COME FROM? WHO WILL NEED THE DATA?

What's up?

The use of the data after the capture process may seriously influence the decisions made before and during data capture. In all cases, data quality should be as optimal as possible given the experimental conditions or the boundaries of the study. Still, data may be generated in some cases for very limited purposes (for instance, the calibration of an instrument) or for massive reuse (sequencing a reference population on biodiversity or climate-change data points) and as said, sometimes for highly regulated and certification purposes. The richness of the metadata needed to enable these post-capture processes must be carefully considered before, and monitored during the actual data capture process.

DO

- Discuss with the PI and the team the exact sources of the data as well as their purpose.

- Point out to the team that they also have to anticipate use beyond the original purpose of the data.

- Capture the richest possible metadata within your experimental and institutional possibilities (it is better to discard some metadata later than to wish you had captured them).

- Clearly divide the experimental results in data that can principally be made FAIR (i.e., also machine-actionable) and those that are intrinsically not machine-readable. Examples of the latter category are, for instance, environmental samples, tissue samples, biological specimens, sound recordings, videos, free text, etc.

- Make a very serious effort to adorn the non-FAIR experimental results (regardless of whether they may be made intrinsically

FAIR later on) with rich FAIR metadata, annotations, and provenance information.

- Make a full list of reagents, sensors, workflows, machines, and other research objects that are involved in the process of data generation.

- Agree with the team about the level of metadata needed for each step in the study process.

- Record the batch and version numbers of all research objects and reagents used in the process.

- Carefully study whether the data capture, storage, and exchange formats have been settled before in previous experiments, either in your team or elsewhere.

- Make a backup of all metadata in digital format (even if they are also written on a tube, for instance), with a PID for every sample and a link between that PID and the actual sample (if the PID can also be sustainably engraved in the sample container, that is obviously the best choice).

DON'T

- Consider experimental results or objects that cannot be made intrinsically findable and actionable for machines to be 'unfair' in the semantic sense. They are key scientific research objects but they are of a nature that makes them not usable for machines (and in many cases humans) without intermediate steps (a tissue sample may reside in a freezer in a plastic tube). However, the metadata describing it will be crucial to make the (set of) samples findable, accessible (with intermediate steps), interoperable (after processing) and, therefore, ultimately reusable by others, including machines.

- Assume that the purpose for which samples or data were originally generated will be the only purpose for which they will ever be used. Instead, try to imagine any other future use to the best of your abilities.

- Separate the (FAIR) metadata any further from the non-FAIR data and samples any further than absolutely necessary to support minimal chance of mix-ups (remember labels falling off tubes in the freezer?).

Resources: `http://dmp.fairdata.solutions/resources/nkj`

4.2   CAPACITY AND HARMONISATION PLANNING

What's up?

Once you know exactly what the output of the experiments or studies will be, both in terms of data types and their volumes, you need to plan for the required capacity to capture, store, process, analyse, and provide them for reuse. These aspects have been covered in general terms before, but here we need to emphasize that it is your task as a data steward to prevent unexpected capacity problems in hardware, but also human capacity that may cause delays or even waste of generated data once the experiments have been set in motion. This does not only cover the capturing of the raw data, although this is the most urgent issue to address. You should also be fully aware of the stability of data and samples generated in each step of the experimental process. In many cases, samples taken from physical or biological systems, for example, have a very limited stability, and measurements on them can be heavily influenced by the time lapse between the time point at which the sample was taken and the time the actual measurement of its features and values took place. Again, this is mainly a part of general good research practice and the co-responsibility of the entire team, but your role as a data steward is to maximally ensure that variations that may be traceable to time-lapse or other experimental conditions are properly recorded and therefore traceable. For instance, the exact position of each sample in the container used for measurements (i.e., a 96-well plate) and the time it took to run the entire plate through the measurement procedure can prove to be extremely important for later processing and analysis of the data. For instance, fluorescence of a marker has a given decay over time, and differences between the first well in the plate and the last one may have to be corrected for that, before the data can be processed further.

DO

- Make sure you fully understand the 'sources of potential variance' that are relevant for the experimental study at hand, and that you advise the research team on which metadata and process-descriptive data have to be captured to minimise risk of data loss or lack of harmonisation, calibration, and correction procedures.

- Capture all reasonably possible metadata and data about laboratory procedures again following the 'better-to-throw-away-than-regret-not-to-have' principle.

- Ensure that correction, harmonisation, and calibration procedures on the data are also considered as part of the experimental process, and therefore monitored and captured as such.

DON'T

- Consider yourself out of the loop during the actual experiment, and see yourself merely as the recipient of the data when they come out of the experimental workflow process.

- Consider the deep knowledge about experimental procedures (not necessarily the hypothesis being tested or generated) none of your business and the realm of the experimentalist, but act as an intrinsic member of the research team during this step of the data cycle.

- Start the actual data capture process without the best possible knowledge about the future use of the data and the requirements to be able to compare them with OPEDAS without cumbersome and error-prone *post hoc* harmonization processes.

Resources: `http://dmp.fairdata.solutions/resources/rqh`

4.2.1 Will you use non-equipment data capture (i.e., questionnaires, free text)?

What's up?

There are a lot of data capture procedures that are not strictly spoken 'instrumental output'. It is obvious that you cannot walk with the researcher in the field who takes water or soil samples, be at the table of the pathologist who takes a tissue sample, or the microscopist who makes sections, and who stains, embeds, and preserves them. Nor can you walk with each research assistant approaching people with questionnaires. However, careful consideration of how data-capture procedures that have a strong non-standardised (human) component such as active questioning of subjects or surgical procedures, may influence the ultimate data you will have to steward is a key task for a good data steward. Just as machines may need to be calibrated at regular times and data may slightly vary based on how long ago that calibration took place, different students may cause different biases in the filling of questionnaires, based on subtle signals they send to the interviewed person. Obviously, as with machines, the data steward cannot change that, but, for instance, careful recording of who took what survey when and under which conditions is extremely important.

DO

- 'Know your source': Make sure that when data are generated in 'variable circumstances by non-standardised agents (such as humans) you are aware of that, and you have procedures in place to detect unintendedly introduced variance (and correct for it if possible).

- Discuss alternative data-view-based concerns with the research team and make sure that all data and sample collectors are optimally aware of the value of standard operating procedures, but also of the need to record any abnormality encountered during the data-capture procedure.

- Keep all records of the capture and pre-processing procedures even after data have been harmonised and normalised. This is not only for review and reproducibility purposes, but also to ensure that if unexpected trends are observed, re-examination of the results can be done with full provenance.

DON'T

- Consider the data or sample capture people and procedures 'none of your business'.

- Leave the considerations about variables in data capture procedures to the experimental team and (again) see yourself merely as the 'recipient of the data' when they come out of the experimental workflow.

- Consider any recorded data and metadata during the capture process as disposable at any later stage unless there is a very good argument to destroy them.

Resources: `http://dmp.fairdata.solutions/resources/ybw`

4.2.1.1 Case report forms?

What's up?

Case reports represent a special kind of data source. Case reports are frequent in medical, biodiversity, and environmental studies, but they can also include case reports of law enforcement officers in daily practice. These case reports are not necessarily always seen by the creator as a research object, but more often as a routine record for internal reporting purposes. However, many of those become research objects at a later stage. In many cases they will be collected by researchers who hope to find patterns, trends, and evidence in such case reports for many different reasons.

In many cases, therefore, these data sources come as they are and you have to deal with them as *intrinsically suboptimal* raw data. Text-mining, as well as human interpretation by others than the creators of the text or the structured data file, can introduce further sources of variance. It is therefore important to carefully prepare the team for such sources of variance. Although some of them are simply unavoidable, maximum care should be taken that wherever this variance can be prevented, it should be. For example, the team should make a clear strategy for how to deal with ambiguous language in case reports, and preferably map everything to the standard vocabulary before analyses are undertaken. Where possible, feedback loops with the original creators of the case reports should be attempted when ambiguities might influence the results. Variance in human interpretation among the team members should be studied and recorded, precision and recall or text-mining programmes run to extract structured information from text should be described and made part of the metadata, and any such sources of variance should be part and parcel of the metadata and process descriptions. Erroneous mapping to concept PIDs by text mining tools is a very frequent source of errors, and should be carefully studied and recorded whenever text mining is part of the analysis pipeline.

DO

- Make the research team fully aware of potential sources of variance and errors in the *post-hoc* interpretation of structured and unstructured sources for research, such as case reports.

- Make the recording and study of non-controllable human behaviour and interpretation a point of study and recording in the broader project.

- Make clear plans with the team about how to correct, normalise, and account for the known sources of variation in such *post hoc* studies.

DON'T

- Treat *post hoc* research objects (not originally being created with a particular study in mind) as equal to research objects, data, and data sources that were purposely and carefully created as data sources for research.

- Ever assume that two people will interpret the same unstructured data source (such as a text) in the same way.

- Underestimate the enormous ambiguity that human-created text contains. Humans are trained to use various synonyms for the same concept (assuming this will keep the text more 'readable' and 'interesting' for others). They are inclined to use jargon, acronyms, and other ambiguous symbols to refer to concepts that are 'obvious' to them in context, but not at all unambiguous for later readers, let alone for machines.

- Assume that machine-interpretation of free text after mining and natural language processing, or of images and audio files, is anywhere near flawless. Lots of controls are needed to reduce the errors induced by machine-interpretation of research objects originally meant for humans.

Resources: `http://dmp.fairdata.solutions/resources/hfg`

CHAPTER 5

Data Cycle Step 4: Data Processing and Curation

In this chapter, the actual processing of data once they are captured or created is addressed. The preprocessing and further processing of data and the associated metadata are core business for the data steward, but so are the workflows, procedures, terminology systems, and formats that will render the data suitable for initial analysis, as well as for reuse in review, reproduction of results, and in combination with other data for future studies, where there is a relatively strong emphasis on the workflows. In many cases, ill-developed or ill-monitored and badly described or unsustainable workflows are being used, which may create severe problems in data interpretation, and induce errors in the next steps in the data stewardship cycle. The challenges regarding data formatting, use of standard formats, and terminology systems are already covered under step 2 of the data stewardship cycle: *Data Design and Planning*, as we believe it is too late to only consider these issues at this stage in the cycle.

5.1 WORKFLOW DEVELOPMENT

General issues about FAIR workflows have been addressed already in the Introduction. Here we will only address issues that are specific to workflows you may run on data for (pre-)processing and curation of data. This is probably the phase in any study where the danger of re-inventing new (slightly faster/better) workflows and algorithms to be 'published in IEEE' is most prominent.

The PI and the entire research team may not be fully aware of the reusable, calibrated, and sustainable workflows and established analytics pipelines out there that effectively process the data type at hand and will produce a standardised and community-compliant output. The chance that someone else has already dealt (repeatedly) with the type of pre-processing and analysis that you want to perform is actually close to 100%. That does not automatically mean that the workflows these other groups have developed are either available to you or that they are exactly fit for your purpose, but in most cases, getting access to them and/or adapting them for your specific purposes is wiser and much more efficient than re-inventing custom workflows for your data processing and curation.

Therefore, this section will mainly deal with the hypothetical situation that workflows exist. If you need to develop workflows from scratch, obviously all considerations for good workflow, standards, and template design pertain.

5.1.1   Will you be running a bulk/routine workflow?

What's up?

A data steward is not a software developer. Developing a new workflow for data processing and curation should be avoided where possible. It is highly unlikely that your data (pre-) processing and curation process is entirely new. Obviously, in the event that the data type is indeed 'new' (the first-of-its-kind instrument, measuring first-of-its-kind data), you will need to develop data processing software that is not co-delivered with the instrument. As said, in that case, follow all recommendations and procedures of software carpentry, realise you are probably producing '*professorware*' in the first place and do not get carried away with the beauty, the publishing and marketing potential of your software. But most of your effort should go into research and consultation on existing tools.

DO

- Check the tool registries of established research consortia or infrastructures in your domain for existing workflows that serve your needs.

- If established registries do not seem to have the tools you need, do a Web search to find less well-known and established tools (that may still save you a lot of time).

- Arrange for remote or face-to-face consultation and training if needed.

- Test-run workflows on reference datasets to get familiar with the system before you submit your valuable new data to it.

- Try to establish contact with the supporter(s) of the software and tools you choose, and make sure you know the version and release policies and practices, as well as the service agreements that are in place.

DON'T

- Assume your data and process is so unique that you need to develop workflows and algorithms from scratch without a thorough (international) search for existing options.

- Present the software or algorithms you may have to develop as 'solutions' that are usable and scalable for others without going through all the steps required to make professional software. That does not mean others cannot use your 'professorware' solution, but be aware of the time and effort it will take to help others to use your tools without proper documentation and support (you will likely be their default 'help desk').

- Ever produce *slightly better* software when good tools are available and their performance is good enough for your purposes.

Resources: `http://dmp.fairdata.solutions/resources/qzt`

5.2 CHOOSE THE WORKFLOW ENGINE

What's up?

In many cases, end-to-end pipelines may exist for the type of data processing you need. However, it may also be true that you need only particular components of a workflow. When workflows are based on serialised components (for instance, Web services) that can be relatively independently deployed, it may be an option to use only part of an existing workflow. However, be aware of the stability issues related to Web services. If Web services are maintained by academic groups, there may very well be issues related to proper documentation, versioning, and support that can severely hamper your experimental procedures. If the reliability of the average Web service is 80% (up-time, versioning etc.) and you need a sequence of five Web services to run your workflow, do the maths on how likely it is that the entire workflow will run smoothly. Therefore, if professional tools exist (even if they are proprietary and not open source), carefully study and discuss the balance between using open source software that you can adapt yourself, its professional support level (which may cost you regardless of the license situation of the software), and the use of commercial software (open source or closed). There is obviously a tendency in academia to use open source software whenever possible, but this does not always add to efficiency, just to the 'perceived freedom to hack', and, is also sometimes based on a strong bias against commercial software and the fear of 'vendor lock-in'. These fears are not always unjustified, but going for unprofessional 'hacks' just to save some money (in the short term) or based on your desire to be able to 'change and tweak' things yourself (or even worse, purely motivated by the desire to publish on something new *per se*) may not best serve your research team. Laboratory analysts would be punished immediately if they crafted their test tubes themselves or made their reagents in a non-Good Laboratory Practice (GLP) environment, so why would that be acceptable for data analysts? So, the choice of workflow engine is as much a team choice (with your advice) as the choice to use reagents such as antibodies from a commercial (guaranteed) provider versus producing them in the lab by immunising mice. Questions to be answered in the process of choosing an existing workflow, engine, or service include considerations about the ease of custom developments. For instance, can a workflow be edited collaboratively, can you reach out to and collaborate with the developers? Does it support the compute-back-ends you need? Does it offer standard tools for the administrative operations? What is the ease of adding new tools? Does it support nesting of workflows? Is it easy to use, for non-computer experts as well; for instance, does it have a running and easy graphical user interface (GUI)?

DO

- Check the various workflow options and engines, not only for promised functionality, but also for:

- Level of documentation, support, licensing, and reliability.

- Keep in mind that your final responsibility is the quality of the processed data, not the creation of new algorithms.

- Get in direct contact with the workflow engine provider and follow training if necessary.

- Run the workflow, engine, and pipeline on exemplar data first.

- Take the 'running' environment into account and check the sustainability of that environment (for instance, Galaxy).

- Check who in the team will have to operate the workflow (experimentalists and ICT people may prefer very different user interfaces).

DON'T

- Develop new workflows or workflow engines unless this is demonstrably unavoidable and consented to by the team.

- Use unsupported tools or environments 'just because they are easy to access and OS'.

- Discard the idea of using commercial or otherwise proprietary workflows without careful consideration and discussion with the rest of the team, especially when your data are difficult to reproduce.

- Expect purely academically published and 'provided' workflows and Web services to be 'up and running' all the time, and consider an entire workflow as one unit, without inspecting all serial elements of it for reliability and usefulness.

Resources: http://dmp.fairdata.solutions/resources/ydj

5.2.1   Who are the customers that use your workflows?

What's up?

Many workflows will have originally been developed by computer or data scientists and scientific programmers. Apart from the first versions usually being professorware (which may still be the best choice you have), they are frequently not very easy for the experimental scientists in the group to use. This means that if you opt for such early development pipelines and work workflows (which again may be your only option in some cases), they might have the embedded consequence that you will have to be on standby when other members of the group need to run the workflow, or in the worst case scenario, that all instances and runs have to be performed by yourself. If the choice of workflow means that you become the 'single point of failure' in the chain of events, this is a sub-optimal choice by default. So, in cases where this choice is unavoidable, make sure you train as many people as possible to use the workflow as well (which again is different from being able to co-develop it in terms of new or customised functionalities).

DO

- Always go for the most 'mature' and user-friendly workflows, those that can be operated by as many people in the research team as possible.

- Consider commercial and proprietary solutions as well, but use the same balancing as described for any choice of workflows.

- Make test runs with different workflows (if available) with the rest of the research team in the loop, and collectively choose the best-suited option.

DON'T

- Choose a workflow that seems easy to you as a data expert, but may be non-operational for other group members.

- Get carried away by options to 'hack on it yourself' at the expense of user-friendliness and stability.

- Overestimate the ability of your fellow group members to deal with difficult or nerdy interfaces (that seem easy and intuitive to you). Many of your team members will panic when they get a strange computer-generated message on their laptop.

Resources: `http://dmp.fairdata.solutions/resources/jrw`

5.2.2   Can workflows be run remotely?

What's up?

If you can choose a cloud-based workflow that can be run remotely, or even better, a workflow that can be containerized and sent to data that remain '*in situ*', this may be the best choice in most cases. As argued before, datasets increasingly become too large or too sensitive (human data) to be pumped around to different storage and compute locations in the consortium you may be operating in. Sending data around is obviously associated with all kinds of technical and security issues. This consideration will be dealt with in more detail when we discuss data analytics, but even for pre-processing and curation of data, moving the data as little as possible and using remote or 'locally downloaded' workflows to deal with these processes is the preferred option.

DO

- Consider the size of the dataset to be processed and/or curated (as far as computer-aided curation is concerned) before you decide to work with a stand-alone, local workflow or a remote service.

- Use workflows that are provided as a 'visiting service' in general as being more reliable and desirable than services for which you need to send your data physically to the compute.

DON'T

- Send data around to services, unless this is the only option or the clearly preferred one.

- Decide on this purely from a data science viewpoint, but, also consult the rest of the team to consider cost issues, sustainability issues, reproducibility issues, and most importantly, ethical and security issues.

Resources: `http://dmp.fairdata.solutions/resources/grt`

5.2.3   Can workflow decay be managed?

What's up?

The stability of workflows over time is a matter of considerable concern and debate (see also generic concerns in the Introduction). In a recent publication [Mayer and Rauber, 2015] it was shown that a serious percentage of workflows collected in a well-established workflow support environment (Taverna) are not re-executable, and often the cause is rather trivial shortcomings, such as lack of example values needed as workflow input parameters, as well as missing libraries for Java programs. The practical consequence of the use of 'academic' workflows, especially those that do not have a frequent use, and often have no service level agreement attached, is that the code is probably also sub-optimal in terms of performance, scalability, and supportability aspects.

DO

- Check the decay risk for each workflow you intend to use before making it part of your routine data processing pipeline.

- Contact the original developers and ask them the pertinent questions about documentation, versioning, support, and sustainability.

- If any of these are unsatisfactory, discuss with the team whether the risks for reproducibility, review, and quality of the downstream data (output from the workflow) are acceptable.

DON'T

- Assume that workflows that are offered on the Web are still the same as when they were first published, even if they have been formally published in informatics journals and are part of a workflow environment, or when they are presented, for instance, in a Galaxy setting.

- Believe the sustainability and support claims for either academic or commercial workflow systems without performing due diligence (including interviewing prior expert users).

- Contribute to the unstable workflow jungle by adding your own professorware solution to the mix without proper consideration of support and decay issues.

Resources: `http://dmp.fairdata.solutions/resources/xyf`

5.2.4   Verify workflows repeatedly on the same data

What's up?

Workflows, especially composite ones, are inherently unstable unless otherwise proven. One way to verify that a workflow is stable, especially if it is composed of multiple independently developed and operating components, is to run it on a regular basis on standard reference and calibration datasets. If results on the reference calibration set are different, apparently one of the components of the workflow is not functioning properly or has been changed. The change may actually be an improvement from the viewpoint of the content provider, but in many cases, updates and new versions are released without proper pre-notification to all users (if these are known and tracked in the first place).

So, even if components of a third-party workflow are improved, they may still cause irreparable damage to your research by introducing unnoticed or otherwise irreparable aberrations or even damage to data. So, especially when a remote workflow or service has not been used for a while in your local setting, you must carefully check and verify that the results of the workflow on your reference dataset are identical to your results the last time you ran the workflow.

DO

- Maintain stable reference and calibration datasets for each workflow operational in your research setting.

- Run these on a regular basis in order to prevent only finding out at the moment that you need it urgently, that the workflow has been changed and your raw data or measurements may suffer.

- Contact all providers of externally-provided workflows to make sure they know you are using their tools, and ask for notification of changes, versions or errors.

DON'T

- Ever rerun a workflow on new data (especially when the results add to accumulating evidence, just assuming that the workflow (even if claimed to be stable) is indeed stable.

- Accumulate evidence and data in collections that were generated at different time points, with potentially changing workflows that have corrupted or otherwise rendered subsets of your collection 'outliers' that may completely mess up the collection.

Resources: `http://dmp.fairdata.solutions/resources/egv`

5.3   WORKFLOW RUNNING

What's up?

A workflow run without supervision is a grave risk to your project. When you run a workflow, you need to have a system in place to monitor progress and potential errors. Not all workflows and services have professional and built-in error messaging, restore possibilities, etc. It is a disaster if you only find out that the workflow has been stalled (a Web service component was off-line, for instance) after a long time, and after potentially irreparable damage to the data has been done. Obviously, always keep a backup of the original data that were entered into the workflow, as some workflows may discard interim files, which may make rescue of data immediately after workflow-internal error only possible if you start all over again with a repaired workflow and the original data.

DO

- Monitor every workflow that is running on a continuous basis (or install/develop machine-monitoring).

- Keep a copy of the original input, to enable restarting of the workflow at any time with the original input.

- Make sure you record every step of the workflow (completed successfully, error detected, component restarted, etc.) in order to be able to trace the source of aberrations in outcome.

- If the input of the workflow is subject to change over time (for instance, sample measurement where quality of sample or staining/labelling is subject to decay), take extra measures (run the supervised workflow with reference data before starting the real experiment).

DON'T

- Let any workflow run completely unsupervised and assume the output is correct and can be fed into the next step of processing or analysis.

- Restart a workflow after a detected error without carefully recording what the error was, and what was done to repair the process.

- Rerun workflows that work on input which may have been subject to change over time (and will therefore give different results when rerun, even if the workflow did a perfect job).

Resources: `http://dmp.fairdata.solutions/resources/dwv`

5.4 TOOLS AND DATA DIRECTORY (FOR THE EXPERIMENT)

What's up?

A full directory of all tools, workflows, and data collections used in your experiments should be available to all members of the research team at all times.

Not only should you avoid becoming a single point of failure in the data processing phase of the research and data cycle, it is also very important and time saving for you to maintain a professional and easily accessible directory of all data processing tools used in each experiment. The best way to do this is to have a general directory where optimal metadata and provenance about the tools and their components is provided. Each tool (component) should be given a unique identifier (including different versions), and for each individual experiment you can refer to this general directory for the tools that have actually been used (and at which point and how) in the experiment at hand. This is especially important for distributed projects. Make sure you run a subset of the workflow/data combinations on all infrastructures to ensure consistency. Making pipelines portable across work spaces (ultimately as VM-type containers) reduces the risk that results will be different in different sub-work-spaces of the distributed team.

DO

- Keep a general directory of all tools used in all experiments and make sure all the team members know of it.

- Make sure that all team members can access the directory and refer to it properly in their lab notebooks and at any other time.

- Ensure that your directory is stable and professional enough to serve as a formal repository of tools that can be used in reviews and reproducibility checks.

- Make a contingency plan for workflow errors, including alternatives (e.g., cluster, grid, cloud).

DON'T

- Run any workflow on any data without referring to its generic representation in the directory.

- Run workflows without recording the exact version used (this can be best approached by having a PID for each version of each tool in the general directory).

- Describe third-party tools in the general directory in your own words without adding sufficient metadata, provenance, and links to allow all members of the team to drill down to the origins of the tool.

Resources: `http://dmp.fairdata.solutions/resources/pzq`

CHAPTER 6

Data Cycle Step 5: Data Linking and 'Integration'

In this chapter, data linking and integration are approached from the perspective of the FAIR principles: How to render data findable (also for others and especially for machines), accessible under well-defined conditions, including basic licensing information, interoperable and therefore reusable. Semantic and syntactic interoperability is explained and the guiding principles are explained, including some hints on how to choose proper formats, workflows, and terminology systems to render data (and tools) FAIR.

Much of data-driven research includes the combination of different data sources, many of them 'born digital'. This process is broadly referred to as 'data integration'. Although this might be philosophically close enough to what is happening, in modern, data-driven science we need to look critically at the implicit connotation of this term. We 'integrate' (elements of) different datasets into one common source of information on which we then base our conclusion. However, in many cases the resulting aggregated information no longer comes from one physically 'integrated' underlying data source (a data warehouse). Increasingly, distributed learning algorithms can visit dispersed datasets, parse them, and manage to obtain the same results, and in addition, distributed large datasets of a relational character increasingly become 'functionally interlinked', rather than 'integrated' in the classical sense.

Major disadvantages of classical 'data integration' relate to the enormous workload needed to create (yet another) special-purpose data warehouse, to then maintain it, and make custom workflows run on that custom-made warehouse.

It may sound counter-intuitive that a distributed learning and compute environment is less error prone, putting much less stress on your internal resources, and will enable the same, or even faster and better, results. Pioneering projects like Open PHACTS[1] have demonstrated that linked data approaches can be cost effective and enable fast results that would otherwise cost many working hours.

The classical approach to 'data integration' is a very logical and in some cases still valid approach: Relevant data may be dispersed over many different databases. In the life sciences, for example, there are over 1600 such databases registered. These come in many different formats, using many different languages, vocabularies, and (metadata) schemas. It is therefore close to impossible today to run distributed learning algorithms over these databases without serious preparatory work. Most of these highly valuable data sources were developed at a time when most of the 'consumers' of the database were exclusively humans. Human searching and reading of databases via GUIs will continue, and the human-readable versions of those core resources should therefore continue to exist and be updated. However, a lack of machine-interoperability of such data resources poses a major problem to machine-learning and distributed analytics involving the data contained in these resources. Therefore, a machine-actionable representation of such databases is a very important goal to achieve in order to enable better data-driven and open science. A relatively new development is the phenomenon of 'linked data'. Maybe the term 'linkable data' is even more appropriate. Linking data in a functional way, so that they effectively serve as a single data source for distributed analytics, is a very powerful way to avoid classical, cumbersome, and error-prone 'extract, transform, load' (ETL) processes over and over again. However, a number of basic principles have to be taken into account. Linked data as such is definitely not enough for data-driven open science. The danger of linking all elements of data together without proper context and provenance is that it makes them close to useless to draw solid conclusions on, which is a drawback of many of the linked open data approaches of the early days. In fact, each minimal 'research object' (for instance, a nanopublication representing the smallest possible association or 'assertion'), should carry rich provenance. For instance,

---

[1] https://www.openphacts.org/

it should give information about under which conditions the asserted association is 'valid', who made the statement, when and whether the assertion has been made by multiple people, whether it was curated, peer-reviewed, and so on. If we take the concept of a nanopublication as the smallest imaginable assertion (in essence, of the type *Subject-Predicate-Object*) and larger research objects as growing aggregations of these assertional building blocks in defined containers (which may contain workflows as well), research objects of increasing size and complexity become the essential 'units' that need to be FAIR for machines. One essence of the FAIR guiding principles is that all concepts to which we refer in the research object should be unambiguously defined and resolved via a unique and persistent identifier. The re-expression of FAIR research objects in human-readable formats will thus also serve the reduction of ambiguity in human language. Therefore, FAIR data and research objects in general will also serve human interpretation.

At this point in the data stewardship cycle, you should refer back to the Introduction where the difference between 'samples', their annotations, metadata, and the 'FAIR representation' of those has been explained in more detail. Obviously, important 'concept categories' in science, such as people (scientists, study subjects) themselves, samples, recordings, etc. escape the direct machine-actionability criteria. Some data elements such as images and recordings may become increasingly machine-readable themselves, but many of these elements cannot be 'integrated' or 'linked' in the sense of this section. That does not make them un-FAIR. On the contrary, linking all kinds of research objects, regardless of whether they can be made 'intrinsically' FAIR should be linkable via FAIR annotations and metadata. For all practical purposes, we also include in this category data that are in human-readable format, potentially 'ambiguous', and thus not 'FAIR', but where metadata exclude misinterpretation. As an extreme example: If a textual record in an electronic health record of a patient contains the notorious acronym PSA. This might introduce a severe homonym problem, even in the defined context of the medical literature (PSA has almost 200 different meanings on the scientific literature). However, if the FAIR metadata representation of the same record refers to the concept prostate specific antigen (KLK3) with the UMLS identifier C1519176, and with a proper URI, properly instructed machines will be able to resolve the meaning of PSA in the underlying data record, and it will be considered 'FAIR'. Based on such metadata, even a text-mining programme will likely assign the correct meaning to the highly ambiguous term in the underlying free text records. Anything in the universe (and beyond) can be the 'subject' of research. We distinguish, for example, raw measurement files, samples in tissue or soil collections, figures, medical records (hand-written or electronic), computer code, satellite images, and actual people themselves, as well as their social media and quantified-self as (sources of) raw data. As soon as any real-life entity becomes a subject of research, it is *also* a source of data. Even when you look at pictures, you will make an instant interpretation of them based on the photons they emit and the conditional knowledge in your brain.

Notably, for each reader, these interpretations and connotations are likely to be different. For that very reason, many of the subjects of research, or data sources, in this realm are not, and will possibly never become, FAIR in the sense of being suitable for direct machine-driven analytics or research. However, as soon as we derive 'data' from each of those subjects that have intrinsic semantics, they have a *meaning* for machines and/or the human mind. Once we record these, ideally they should be 'FAIR-borne'. For example, the annotations and the metadata of the 'sample in a plastic tube' should ideally be FAIR. Also, interpretations of a figure and the satellite image as well as their annotations and metadata, should be FAIR. If we refer to a researcher, we should ideally use her or his ORCID and not call them by their given names. These FAIR assertions about intrinsically non-machine-readable objects are in fact all a form of annotations and metadata.

In an ideal world, 'FAIRification' of data and metadata should start as early as possible in the data life cycle. This is obviously not the case with most of the valuable and reusable data we have gathered in the many decades behind us. There is a plethora of data types, ranging from plain text (a nightmare for computers) to all kinds of 'structured data', again ranging from simple spreadsheets to complicated relational databases. However, even in relational databases, many records contain textual and therefore ambiguous symbols. This unfortunate situation makes it exceedingly difficult to search across databases. Multiple synonyms are used for the same concept, and homonyms cause ambiguity. Therefore, false positive results are generated. Quite literally *on top* of that, many of these valuable data sources have their own (search) GUI, non-interoperable APIs and so on. It should be clear that this situation cannot be solved by a top-down standard-setting exercise, nor by creating a global integrated data warehouse, nor by, for instance, forcing all hospitals in the world to use the same, or at least intrinsically interoperable, electronic health record. Most hospitals may have people in their psychiatry ward who have tried to do just that.

So, the crucial element to consider when you publish data that 'can' be principally made FAIR is that as low in the stack as you can afford, you make data or metadata FAIR. But you also need to make clear that certain data sources cannot be FAIR, and that does not make them less valuable. If they are found because of FAIR metadata (for instance, an expert's knowledge profile, based on her ORCID and the concepts she published about, or a sample based on its FAIR annotations), these sources become part of the FAIR ecosystem as envisioned in the Internet of FAIR data and services.

6.1   WHAT APPROACH WILL YOU USE FOR DATA INTEGRATION?

The approach you choose to combine datasets for analytical or other purposes should be driven by the answers you want to get from these data. This may sound like a 'truism', but in fact you may choose quite different approaches for data linking and/or physical integration, depending on your goals. Many compute jobs may be parallelised, and can be run simultaneously on different computers, each on a subset of the total range of data sources. An example is the Personal Health Train [2], where lightweight workflows visit data at mobile devices of people from all over the globe to collect conclusions, while never taking any data out of the original FAIR stations. The other extreme may be that you need to bring data physically together in a high-performance, centralised computer facility to answer your question. In some cases, the 'linkability' and cross-mapping of data may be critical, while in other cases you may still have to choose among costly and cumbersome ETL approaches to create a new data warehouse for centralised computing. Most analytics jobs will fall in the wide spectrum between these extremes.

So, there is no 'one-size-fits-all' solution to data integration and linking. In some cases, linked data may be the solution; in other cases that approach may utterly fail. The general rule, however, is to clearly separate raw data and 'principally non-machine-actionable research subjects' from semantically operable preprocessed data and information. FAIR principles pertain to the latter, but as stated, by subjecting

---

[2]www.personalhealthtrain.nl

semantically operational metadata and annotations to research objects, they nevertheless become part of the FAIR ecosystem. However, some analytical procedures will not need semantically operational data at all, and will just discover meaningful patterns: in raw sensor outputs, for instance.

So, again, dogmatism about how data should be handled is an impediment to scientific progress.

In any case, data should be 'published' (made re-usable for others than those who generated the data) in such a way that they become optimally findable, accessible, interoperable, and reusable for others. Sometimes this will mean linked data and sometimes completely different data formats.

6.2 WILL YOU MAKE YOUR OUTPUT SEMANTICALLY INTEROPERABLE?

What's up?

In case you decide that the data type you want to link or integrate is 'semantic' in character, the linked data approach (regardless of the schema and format chosen) should be maximally guided by the FAIR principles. One frequent misconception is that FAIR is the same as linked data, and linked data is the same as 'RDF' or 'semantic Web'. The FAIR principles do not demand any of that, they simply ask of you that you take optimal care of the four elements of the FAIR principles (for full context, see http://www.nature.com/articles/sdata201618). Please recognise once more before studying the principles below that they may only pertain to the metadata or the annotations of otherwise non-machine-recognisable or -actionable objects. In that case, a unique and persistent identifier should be assigned to the physical object (a tube, a piece of text, a picture, a person) and the metadata should be persistently linked to the physical object, so that it becomes findable and accessible for reuse, even if there may be several steps in between before the actual data become interoperable, linkable, or integrated for the study you want to use them for. This may even include reanalysis, further measurements (looking for new metabolites that were not measured before in an old sample), but the fact that the object that can be a relevant source of data is available under well-defined conditions for reuse in research should be adequately described in the metadata and annotations associated with the object.

To be Findable:

F1. (meta)data are assigned a globally unique and persistent identifier

F2. data are described with rich metadata (defined by R1 below)

F3. metadata clearly and explicitly include the identifier of the data it describes

F4. (meta)data are registered or indexed in a searchable resource

To be Accessible:

A1. (meta)data are retrievable by their identifier using a standardised communications protocol

A1.1 the protocol is open, free, and universally implementable

A1.2 the protocol allows for an authentication and authorisation procedure, where necessary

A2. metadata are accessible, even when the data are no longer available

To be Interoperable:

I1. (meta)data use a formal, accessible, shared, and broadly applicable language for knowledge representation

I2. (meta)data use vocabularies that follow FAIR principles

I3. (meta)data include qualified references to other (meta)data

To be Reusable:

R1. meta(data) are richly described with a plurality of accurate and relevant attributes

R1.1. (meta)data are released with a clear and accessible data usage license

R1.2. (meta)data are associated with detailed provenance

R1.3. (meta)data meet domain-relevant community standards

DO

- Make a careful analysis of the objects and data that can be made FAIR and those that can only be made part of the FAIR ecosystem through semantic annotations and metadata.

- Create the metadata and identifiers according to the FAIR principles, regardless of whether or not the data or the object/data source itself can (also) be made FAIR.

- Follow the FAIR principles for all data types that can be made machine actionable.

- Consider the guiding principles in every step, for instance, (R1.3), by choosing a vocabulary that meets (or is properly mapped to) domain-relevant community standards.

- Always keep in mind that the end goal of FAIR data creation and publishing is optimal reuse for your own and others' studies that require functional interlinking or integration of multiple datasets.

- Keep and publish annotations on existing datasets that you reused and add them to the FAIR ecosystem by exposing them in a FAIRpoint or a FAIRport[3] even if you are not the original creator of that data.

- Study and implement where possible the 'Joint Data Citation principles' in order to make your data maximally reusable and citable, so that your institution can do metrics on and reward you for offering your data and annotation for reuse by others.

DON'T

- Try to make data or other research objects FAIR if they are intrinsically not machine-actionable.

- Make data machine-actionable without considering a proper (and preferably unambiguous, human-readable symbol) for each concept; URIs are as annoying to people as human language is to computers.

- Allow any concept-denoting symbol in your (meta)data that is 'ambiguous' and will therefore frustrate both *in silico* and *in cerebro* reuse by others.

Resources: `http://dmp.fairdata.solutions/resources/fxm`

---

[3]See for instance: https://www.dtls.nl/fair-data/find-fair-data-tools/

6.3 WILL YOU USE A WORKFLOW OR TOOLS FOR DATABASE ACCESS OR CONVERSION?

What's up?

Workflows for FAIR data handling are increasingly available. For most data types and metadata needs, there will be earlier examples, emerging community standards, and publication, as well as linking, integration, and analytics workflows that can handle FAIR data.

As of 2014, the year that FAIR as a principle was conceived, a growing series of so-called 'Bring Your Own Data' workshops have been organised, where data owners of various basic data types have worked with FAIR data experts to represent their original data in a FAIR manner at the metadata level, and, in many cases also to turn (part of) of their core data into FAIR format. In most cases this procedure did not *replace* their original database, but it merely exposed part of their data in a FAIRpoint, meaning that FAIR-aware tools will find them and will be able to operate in that subset of the data. By the time you read this book, many data types will have been 'done before' and an increasing number of concepts you will have to refer to in your (meta)data will be covered by community compliant — and — supported vocabularies. There are international efforts and research infrastructures in most domains that are in full swing to develop tools, standards, and best practices for their domain, where these do not already exist. Granted, there may be large differences in maturity of data-type registries, proper, controlled vocabularies, mapping services between these terminology systems between the scientific domains. However, in most science domains, increasingly, front-line research will have to link and meta-analyse data from different domains and disciplines. For that process to be effective and reproducible, it is imperative for interdisciplinary and open science that data 'talk optimally' to other data within your domain, and also to data from other domains. Therefore, your search for 'pre-existing best practices, formats and standards' should not stop at your disciplinary border. One of the best approaches to getting acquainted with best practices of the data type you have at hand is to turn to established research infrastructures in your domain, or consult certified data stewardship nodes in your domain.

DO

- Consult experts, preferably associated with established research infrastructures in your domain.

- Also consult outside your domain (there may be a vocabulary in agriscience you may want to extend for a biodiversity project, or one in meteorology you may be able to use for part of your geology data, as well as one in chemistry that may cover the metabolites you want to refer to).

- Detect concepts or formats that you need and for which no appropriate 'prior art' is findable or reusable in your specific case.

- Always approach research infrastructures or services (like ELIXIR, NCBO, or EPOS) for advice. In many cases, customisation of existing data models and formats or extension of leading ontologies will be the best option to cover these identified gaps.

DON'T

- Invent new data types, linking, publication, or integration pipelines, unless you have exhausted all possibilities to use or customise existing ones.

- Create new identifiers for any concept in your dataset unless the concept is nowhere defined with the exact defined meaning you want to attach to it. If you really need to create a new identifier for a new concept (for instance, a phenomenon or a species you describe for the first time in the scientific discourse), always try the route of extending community-preferred vocabularies and ontologies first.

- Link and integrate datasets that are intrinsically non-harmonised and incomparable or incompatible. Some inconsistencies will automatically be detected by computer programmes but the 'crap-in, crap-out' mantra is also true for the meta-analysis of linked or integrated data.

Resources: `http://dmp.fairdata.solutions/resources/qqb`

Data Cycle Step 6: Data Analysis, Interpretation

Models in the strict sense are not really covered in this book, but here are a few words to explain why. First of all, the term 'model' may be used in many different ways (even within the science realm) as much as the term 'ontology'. One could define models as any representation of a physical object (even, for instance, a scale model of a car), but in science, models are in general a representation of a hypothesis on how concepts relate to each other. Within that broad definition any representation of knowledge (down to a single assertion of the type subject-predicate-object) could be regarded as a 'model' of a piece of knowledge. However, such a broad definition would not be very useful in the context of this book, as it would in basic reflection of the science of knowledge modelling.

We will therefore define models here as knowledge representations that are 1: **'hypothetical'** in the sense that they are being 'tested' and 'adapted' to best fit the phenomena we observe and try to model, and 2: **'dynamic'** in that they also model a 'process' rather than just a piece of static knowledge.

Many models are tested in terms of 'predictive value'; for instance, climate models or physiological models trying to capture and predict metabolic processes. In some cases, there is a fine line between models and less dynamic representations. For instance, is a classical biological 'pathway' a model? It could be argued both ways: It does not fit the more narrow definition of the 'models' that we decided to exclude from this book. Although a pathway is a 'static snapshot' of a process and is in many cases at least in part still hypothetical, the pathway representation as such does not necessarily describe the dynamics in the actual process the pathway represents. If, however, the pathway contains the differential equations that describe the transition from one state of a node in the pathway to another (for example, an enzymatic process), and can thus be used to test whether it reasonably describes what is really happening based on variable input and output parameters, it would become a 'model' in the narrower definition.

So, whenever we use the term 'model' here, we mean 'data model' rather than a testable 'model of reality'.

7.1   WILL YOU USE STATIC OR DYNAMIC (SYSTEMS) MODELS?

What's up?

Each set of raw data, but also more processed (relational) data and information, is captured in some sort of data 'model'. We distinguish data 'model' from data 'format' as the format is really just a 'form' in which data are presented, such as CSV, Matrix, RDF etc. However, it is very important for a data steward to carefully choose a data 'model' in the sense that the data are presented in the most meaningful and easily understandable (and thus reusable) manner. In particular, the model should be machine-readable and -actionable in nature. This goes beyond the formatting and identifier-related consistency of the data themselves. The fact that we handle this issue under the interpretation and analysis section indicates that this has everything to do with what the data will be used for. It might be that for a particular study, data have to be remodelled, including, in some cases, changing them to another data format, or, for example, changing the units of measurement. This again emphasises the enormous importance of proper annotation (which units of measurement were used, etc.) and provenance of the dataset. Future users of the data may use them for very different purposes than those that were the reason to create them in the first place.

DO

- Carefully choose the most appropriate data model for your data.

- Describe data model choice, units of measurement, vocabularies used, and so forth, in great detail in the metadata.

- Include in the metadata issues that may relate to conversion of (elements of) the data into other models or formats.

- Carefully study the data (especially OPEDAS) before assuming that you can indeed use them for integrated analysis with other data.

DON'T

- Ever create a new data model or format before carefully checking that no suitable format exists.

- Assume that the nature (and format choices) of elements of your data (such as units of measurements) are 'self-evident' so that you do not have to specify them.

- Integrate data for analysis without careful consideration of the compatibility and interoperability of the different data models that 'enter your analysis workflow'.

Resources: http://dmp.fairdata.solutions/resources/ykv

7.2   MACHINE-LEARNING?

What's up?

In case your analysis will be (partly) based on machine-learning, you need to be aware that machine-learning algorithms principally 'build their own model' based on 'exemplar' inputs, exploit algorithms that 'learn' from those examples, and then exploit the 'learned' models to predict other outputs, and, for instance, discern similar patterns in novel data. These algorithms are very different from those that 'know the data type they are running over' and therefore are strictly programmed to follow particular instructions reproducibly. Machine-learning is usually employed in computing tasks where designing and programming explicit algorithms is infeasible. This obviously has an impact on the required quality of the data. Intuitively, you may think that the 'machine will figure out' what it is looking at, and that the FAIR principles do not so much apply to the type of data that are used as input for machine-learning. Granted, the data elements themselves may be 'unstructured' (for instance, an x-ray image) and thus intrinsically not FAIR. However, in this case, the quality and FAIR elements of the annotations as part of the metadata may be even more critical than in cases where the data elements themselves are directly machine-actionable and 'understandable' by the machine. For instance, if an algorithm is designed to learn from patterns and colours in images, proper context as to in what category the image should be placed is extremely important and should be part of the machine-actionable metadata of the image. So again, clearly distinguish between the three categories of data defined earlier: Intrinsic metadata, user-defined metadata, and the data elements (including the 'raw' data themselves). Machine-learning may take place both at the metadata and the data-element level.

DO

- Distinguish the metadata clearly from the data themselves (machine-learning may take place on both).

- Make sure it is easy for analytics workflows to either 'include' or 'exclude' metadata and annotations.

- Clearly indicate in the metadata what file type and data model the actual data is in, and how it can be accessed for machine-learning purposes.

DON'T

- Put the metadata and the data elements in the same file without a clear distinction (machine-learning should be possible on both or separately).

- Separate the actual data from its metadata to such an extent that the learning algorithm may find the metadata, but has great difficulties, without human intervention, running on the 'raw' data they describe. Even if you consider the raw (source) data not

FAIR, as in 'immediately machine-actionable', they may be actionable in the specific case of this algorithm (from text-mining to any other pattern recognition algorithm that works on unstructured data).

Resources: `http://dmp.fairdata.solutions/resources/wgj`

7.3 WILL YOU BE BUILDING KINETIC MODELS?

What's up?

In kinetic models, the spatio-temporal elements of the phenomenon under study are critical. This means that data to be used in such models need to have the richest possible time interval, and space coordinates attached to them. More so than in data that will be used only for static hypothesis, pathway, or ontological model purposes, data to be fed to kinetic models must be adorned with rich provenance and metadata about space and time aspects. Obviously, there might be many datasets that were not originally meant to feed kinetic models, but can be used for those later on. It is therefore wise to consider all spatio-temporal parameters to be of potential importance for reuse of your data, regardless of whether they were purposely created for kinetic modelling. It should be clear that 'models' as such are also research objects, so all FAIR principles pertain to them, as well as the rule that you should never make a new model if there are existing ones available that suit your purpose.

DO

- Always ask the research team whether kinetic modelling will be an option in the data analytics steps of the study.

- Make it a habit to collect and attach 'richer metadata than you would imagine necessary', as not *using* them will be an option, but not *having* them might prove irreparable later.

- Take extra care of capturing the most rich parameters regarding space, time (-interval), volume, level, etc., in case data are purposely generated to function in kinetic modelling studies.

DON'T

- Lightly assume your data will never be used for kinetic modelling.

- Fail to discuss the options that this might happen with your team.

- Underestimate the 'granularity' of the needed time, space, and level details to be recorded.

Resources: `http://dmp.fairdata.solutions/resources/hjq`

7.4 HOW WILL YOU MAKE SURE THE ANALYSIS IS BEST SUITED TO ANSWER YOUR BIOLOGICAL QUESTION?

What's up?

Choosing the best analytical approaches and tools is largely outside the scope of this book, as it is at the essence of data analytics itself. However, there is a data stewardship edge to this question as well. First of all, it is obviously very important to discuss in step 1, with the best statistician you have access to, the nature and the (sample) size of the data that you need to answer the question pertaining to the newly generated data for the experiment at hand. A very important data stewardship aspect is that in data-driven research, very frequently, OPEDAS and multiple novel datasets need to be collectively analysed to get the sought-for answers. This first of all means that you can never meaningfully address the question of required sample size and scope of the new data generation effort without having carefully explored the existing reusable data that may be useful to answer your question. So, next to the 'data-intrinsic' aspects of size, scope, and quality of your datasets, and the experimental-data versus control-data aspects, there is a serious challenge in functional integration or linking of your new data with other data. The size, format, richness (also of metadata), and the data model chosen for the capture and storage of your data may be influenced by the other data that are needed to get to a final interpretation of the patterns and phenomena you want to study.

DO

- Consider the 'functional integration and linking needs' of newly generated data with the team in step 1.

- Choose wherever possible the same data formats, metadata standards, vocabularies, and data models for your new data as those that were used for the data with which your data should interact.

- Consider and provide for the mapping, parsing, and transformation jobs that may be needed to put other data in a format that can make them functionally linked and interoperable with your data. An important aspect of FAIR data is that machines should, wherever possible, be able to parse or otherwise transform any data into any other suitable format (requested by a particular tool, for instance) with zero or minimal errors.

DON'T

- Consider data as a one-off asset for a single experiment (most data will be reused inside or outside their original scope/study).

- Limit your metadata to the conceived need for the study for which the data were generated, but collect as many as are achievable and potentially meaningful.

Resources: `http://dmp.fairdata.solutions/resources/grn`

7.5   HOW WILL YOU ENSURE REPRODUCIBILITY?

Reproducibility of data is a major issue in contemporary science. There are many reasons why study results may not be exactly reproducible. It is very important to distinguish fundamentally flawed experiments, studies, and results (and the reasons why these are not reproducible by others) from results that are not (exactly) reproducible, but still support the generalised conclusions drawn in the original study that produced them. There is a lot of literature about this fundamental methodological aspect, and it is strongly recommended to read up about it, so that you are properly aware of all aspects related to reproducibility of experimental conditions. However, what you are mainly responsible for as a data steward is the actual 'intrinsic' reproducibility capacity of your data. This again means that their metadata should always be as rich and complete as possible, and they should be FAIR. It is also well documented that human reading, transformation, and interpretation of text and data files is error prone, and that the consensus of human readers and observers of the same dataset or text rarely fully align. As a matter of principle, however, identical machine workflows, running on the same identical data, should give identical results. Elsewhere, we have already listed many reasons related to data capture that may lead to difficulties in reproducing the same study or experiment later on. Here we address the 'intrinsic reproducibility value of the data' after they have been captured, preprocessed and stored for reuse. Obviously, the documentation and versioning issues described for data capture are as crucial in this phase as they were in the capture phase. Mortal sins are, naturally, the non-documentation of 'what you did with the data after capture', before they became what they are in their stored form, and the storage of samples, as well as their annotation with metadata in a format that may be subject to unintentional change, decay, or that may soon become unreadable. Paradoxically, the fast developments in formatting, data carriers, and readers have the direct consequence that papyrus rolls of thousands of years ago can still be read with the same eyes and the same brains as when they were created, while even floppy disks (remember?) and VHS tapes need 'old machinery' to be read. Reusability of data (and therefore, reproducibility of the study they were generated for) obviously starts with 'finding' them in the first place. Next, they need to be accessible (under well-defined conditions, so that machines as well as people know that they are 'allowed' to reproduce the data, and they should be understandable, and therefore interoperable, with other data. Obviously, reuse of data is much broader than just using them in a peer-review process or otherwise for reproducibility, but for this key process in science, data clearly should also be stored and published according to the FAIR guidelines.

7.5.1 Will this step need significant storage and compute capacity?

What's up?

The processing and storage of data is no longer a trivial issue as far as compute and storage capacity is concerned. Later we will explain the vast difference between data 'archiving', 'publishing', and offering data in 'high-performance analytics environments'. For now, it is important to realise that turning vast datasets into their final processed form, publishing them, and offering them for effective reuse may put significant burdens on your ICT and expertise. So, this is an aspect to consider already in step 1 of the data stewardship cycle, during the design of the experiments.

DO

- Consider data size and complexity of the data to be generated in the experimental design process.

- Include the perceived needs for data archiving, but also internal and external reuse issues in the capacity planning.

- Clearly distinguish between 'first-off' use of the data for the study and the needs of peer reviewers to reuse the data for reproducibility checks only, as well as the wider community that may want to use your data as OPEDAS for entirely different purposes.

- Plan and budget for both, as just providing your data for reproducibility checks will usually not be enough for a good data stewardship plan.

DON'T

- Underestimate the time and resources you need to make the data FAIR, for reproducibility and other reuse purposes.

- Run processes leading to subsequent stages and formats of preprocessed data on unspecified and 'local machines' and with undefined software.

Resources: `http://dmp.fairdata.solutions/resources/wqm`

7.5.2   How are you going to interpret your data?

What's up?

Data *interpretation* leading to domain-specific scientific hypotheses, conclusions, and applications, is not necessarily a core expertise of each data steward. However, basic knowledge of the domains relevant for the interpretation of the data is critical. A basic understanding of what the 'domain experts' (increasingly teams composed of experts from many domains) study is crucial for a good data steward. The formats, and the needs for 'immediate availability' status of data, may be seriously influenced by the choices of the research team and, especially, the way in which they want to go about interpretation. If interpretation should mainly be done by human reading and reasoning, the requirements for your data output and format will obviously be very different from massive graph traversing to reveal connectivity patterns, followed by conformational, human reading in externally linked literature and databases. Not only that, but when interim results clearly indicate major aberrations that are 'simply impossible', you need to be able to spot these before the next working meeting of the entire group reveals them or is severely hampered by obviously faulty outcomes.

DO

- Consider the anticipated methods for data interpretation very early on (step 1), as they will influence many other steps in data capture, preprocessing, curation, and functional integration.

- Make sure you have a basic understanding of the domain knowledge 'around the table' in the research team.

- Check all 'intermediate outputs' in the process of data processing, linking, and analysis for obvious gross anomalies that may indicate errors in parameter setting, etc.

- Carefully record all steps leading to and followed in data interpretation procedures (including where human error and bias may come in).

DON'T

- Act as if you only come in when the data are there, process them nicely, and deliver beautiful graphs and pictures.

- Try to become a semi-domain expert: You should leave the final interpretation of the results to the experts in the team. Especially when there are multiple domains involved, being expert in all these domains is impossible, and trying it anyway may be counter-productive as well as irritating to the real experts.

Resources: `http://dmp.fairdata.solutions/resources/hfj`

7.5.3 How will you document your interpretation steps?

What's up?

Here we can be brief. All issues pertaining to documentation and provenance of the earlier phases (data capture and processing) also pertain to this phase in the data stewardship cycle. The only thing that may be different is that during this phase, it is unavoidable that human input, and therefore error and bias, will enter the game. It is therefore the task of a good data steward to carefully follow (and watch) the interpretation process, and warn the team if variation in human interpretation of the actual data will cause aberrations, especially if the interpretation steps involved data curation, selection of subsets of data, and reading of unstructured text. Again, recording as much as possible what happens is key to good data stewardship.

DO

- Act as a 'methodological watchdog' during the data analytics and interpretation process and record 'everything going on', including unexpected changes in interpretation.

- Make sure that all players in the interpretation process are fully aware of the *critical points* in the evaluation of the results and the data where human bias and error (that you cannot control at the data and the machine-workflow level) may influence outcomes.

- Encourage everyone to record very carefully what happened exactly at these points, especially when the data are curated or turned into another format.

DON'T

- Assume that people are as 'rigorously reproducible' as machines and workflows (in the ideal situation).

- Interfere with the interpretation process itself unless there is a clear misunderstanding of (the nature of) the data.

- Step out of the interpretation process altogether, as you will be unable to 'track back' major errors that led to misinterpretation and irreproducibility of the conclusions.

Resources: `http://dmp.fairdata.solutions/resources/gwf`

7.6 WILL YOU BE DOING (AUTOMATED) KNOWLEDGE-DISCOVERY?

What's up?

If the interpretation process did involve automated querying of reference resources and/or, for instance, text-mining pipelines, you should be aware, and make the rest of the team aware, of all issues pertaining to using OPEDAS described earlier. OPEDAS and workflows may be subject to versioning, extension, curation, etc., which may heavily influence sequential results of such processes, even if they are automated. Especially when such query and interpretation pipelines are used by multiple members of the research team, they should all be working with *exactly the same version* of all tools and resources. This can be achieved much more easily if all interpreters are working on standardised machines you provide, but in most cases, teams are spread around the world and may use either online versions of tools, Web services or locally installed versions, or a combination of all these. If these are not calibrated and aligned in the same scrutinised way as the machines that captured the data in the first place, major errors may be introduced.

DO

- Make the research team fully aware of all error sources imaginable during the interpretation process.

- Be fully aware of each and every person in the team who plays a role in the interpretation process.

- Ensure that all of them have EXACTLY the same version of every tool and resource they need for the interpretation process.

- Ensure that all 'automated discovery' pipelines that may be started up during the interoperation process are identical.

- Log every step of the interpretation process and force the domain specialists to record versions of every tool and resource they use.

- Record with each subset of interpretation-related datasets who it was that created that interim set, or subset.

DON'T

- Trust that domain specialists are fully aware of the many pitfalls in tools that come across to them as 'commodities', and provided by the data expert.

- Contribute to the messy situation by being sloppy in version control of tools and resources yourself.

- Pool (interim) interpretations from different people in the team or from different automated systems before checking the integrity and the versioning of everything and every step.

Resources: `http://dmp.fairdata.solutions/resources/bzu`

CHAPTER 8

Data Cycle Step 7: Information and Insight Publishing

As discussed in the Introduction and in the open science animation[1], we clearly distinguish between the publishing of (narrative) conclusions, claims, and insights in classical scientific articles and the publishing of data. Narrative (not necessarily journals as we know them) will always play a role in the *digging* part of open science. Computers and computer-readable language are not very well suited to interpret human language and associative reasoning. So, the description of why and how the study was done, how the results were obtained and why the claims based on the data are valid, will always need a human-readable element. Rhetoric and argumentative language are largely meant for people, and although progress is made in structuring arguments and provenance trails in more structured ways, we will for now assume that these are still part of the narrative part of science. Machines will for now play a minor role in this part, and the 'digging' tools in text and classical databases are very different from the high-performance analytics tools that will use your data. Obviously, from a philosophical standpoint, narrative articles are research objects, and they are also 'data' in a certain sense, but narrative is intrinsically difficult to read for machines and therefore we generally do not consider unstructured and structured text fields FAIR for machines. So, although they played

---

[1]https://www.dtls.nl/5825-2/

a critical role in science and will continue to do so in open and data-driven science, we will not address the 'writing of good articles' here. Still, as articles are research objects, too, the metadata of any article your group produces should be FAIR, so that computers can at least find it. Ideally, all concepts referred to in your texts as well as all scientific claims made that *can* be added as metadata to the text in FAIR format should be added. However, as also discussed in the Introduction, textual search will miss the vast majority of data and information in the deep Web. It is therefore crucial that search engines can find your data, *independently* from the article that was written in conjunction with them. If the only access to your data is their presence in a TIF file, as a table, or via a supplementary data link to an unreadable PDF, your data are simply not part of the open science ecosystem and will not be reused effectively.

Therefore: A good data steward never publishes data 'hidden behind an article'. A good data steward publishes data in FAIR format and adds a 'supplementary article' to make the data understandable and reusable for people as well.

Data publishing should follow FAIR guiding principles, and we refer to external resources and tooling for how to make sure the findability, the accessibility, the interoperability and therefore the reusability of your data is optimal. Findability is largely related to quality of the data and particularly the metadata, as well as to the online findability of the repository in which they are actually stored. Therefore, the general rule is to make sure that any data you publish deserve the FAIR qualification, and that they are published in a trusted repository. In the near future, FAIR metrics and certificates for FAIR-compliant services are expected and those will be accessible through the data stewardship wizard[2] as soon as they become available.

8.1 HOW MUCH WILL BE OPEN DATA/ACCESS?

What's up?

Open access is a term frequently associated with 'science 1.0', when scholarly communication was still largely done via academic journals, later *extended* to the co-publication of 'supplementary data'. The actual article text did contain the new claims and the argumentation

---

[2]https://dmp.fairdata.solutions/

around them. This publishing system slowly became perverse as text was hidden behind firewalls and then the reader had to pay. Open access as we know it is just another business model where the 'polluter pays', in other words, now the author pays a publication fee and the text is free to read for everyone. Much as this is a general improvement, it does not fundamentally support open science better than the old system. First, authors in less fortunate institutions cannot easily pay for publication, and are once again put in a disadvantaged position. Second, there was already far too much to read anyway, and now there is even more to read. Third, data, the key asset for open science, is not necessarily more findable in open access papers than in closed papers (the TIFF walls and the broken links as described in the Introduction are not opened up). Therefore, *publishing data in their own right* in open access is extremely important, and a key task for a good data steward. We also need to clearly distinguish here between metadata and the actual data. It can be argued that metadata of any research object, regardless of whether it is an article, a dataset, or a piece of software should be FAIR and open access (the A in FAIR is distinct from open as was discussed in the Introduction). There might be very good reasons to keep actual text or datasets under well-defined, but restricted access. This may have roots in patient privacy, or national security considerations, but also, especially for data generated in the private sector, reasons of competitive advantage. However, the latter is very difficult to argue for data that are generated with public funding. They should be considered a public good by default, and reasons to keep them from reuse by others will have to be argued in any public data stewardship plan. Arguments to separate massive and crude data from FAIR metadata have also been discussed earlier. The raw data may be too large to keep in a high performance reusability (HPR) environment, which is typically orders of magnitude more costly than, for instance, on tape. So, in general, FAIR metadata allow your data and tools to participate in the FAIR open science ecosystem.

DO

- Clearly consider for each dataset whether it is regarded as being generated by public or by private funding, or a mixture of the two.

- Take as the default that metadata are always FAIR, with the accessibility level being open access without any restrictions.

- Make a conscious decision with the research team for the best license to put on any given dataset (the actual data elements).

- Include the intended level of restriction in your data stewardship plan (even at the proposal stage). Again, for most funders the default will be open, so in case you want to argue for a more restrictive license, prepare your arguments carefully, as they may influence your chance to get the proposal awarded.

- Work wherever possible with open source software, but never at the expense of aspects of professionalism, versioning, documentation, sustainability, and (SLA) support.

- When dealing with software, make your code open source under a well-defined license, but consider sustainability issues before doing so. Open source code without proper community or licensed support is a 'poisoned gift' to the community as it may impair reproducibility of research in a serious way.

Resources: `http://dmp.fairdata.solutions/resources/jvm`

DON'T

- Consider data or software without a proper license open and 'reusable'. It is not, and especially companies will most likely not touch data or software until the licensing and support situation is clear.

- Publish data or software with a license that is any more restrictive than what you really need.

- Use restrictive licenses on data or software unless you really need to, for privacy, security, or sustainability reasons. These will have to be argued for in future data stewardship plans.

- Call data 'open' and 'reusable' unless you have made sure they are also actually findable outside your group, accessible under well-defined conditions and interoperable for machines with only trivial adaptations. If any of those three elements is missing, they are still 'reuseless'.

- Publish data or software without the attributes and the instructions that make them properly citable.

Resources: `http://dmp.fairdata.solutions/resources/cwq`

8.2 WHO WILL PAY FOR OPEN ACCESS DATA PUBLISHING?

What's up?

Open access publishing of articles is now an eligible cost on most grants and increasingly, institutional policy is to support the publishing of articles in open access journals. However, the actual FAIR publishing of data is not yet as obvious. The example of the FANTOM5 paper in the Introduction is an instructive example of how data that are associated with a paper can be a very important asset by themselves, but will also cost significant time and effort to be published in FAIR format (in this case nanopublications). If we stick to the principle that data needs to be published in FAIR-compliant formats to be optimally reusable, separate planning for the capacity and the resources to do so is part of good data stewardship. Publishing an open access article with the actual data still being hardly findable (i.e., hidden in supplementary data files on the deep Web or without proper, dedicated metadata), accessible only in non-computer readable format, and/or without a proper license. Interoperable means that there are no ambiguous, textual or non-community adopted identifiers, which renders data very difficult to reuse. Therefore the data on which a paper is based should be published in their own right and the 'supplementary article' (preferably open access) should be regarded as the description of the rationale, the methodology of the study that generated the data, and the *first* set of conclusions and claims based on the paper. Returning to the FANTOM5 paper, the number of derivative papers based on the same dataset effectively shows the 'bankruptcy' of the classical publication method. The dataset is key and all papers (including the first 'mothership' paper) should be seen as derivatives.

Now, the question arises, as to how much of your data can be open access. As argued before, open science requires open access as the default approach. However, there are situations where the data cannot be made available in open access without restriction, even if the first and all subsequent derivative textual publications can be open access. For instance, data that are national security sensitive, or privacy sensitive, such as patient data, may have to be restricted in access. Obviously, this already poses a fundamental issue for peer review, as the reviewers would have to 'trust' that the conclusions drawn in a particular derivative paper are based on solid data, and represent credible conclusions. Therefore, even if raw data are to be restricted, a good data steward will always do everything needed and possible to provide a dataset (anonymized, for instance, or aggregated) that allows the best possible review, reproduction and openness. In case the underlying raw data have to be restricted the provenance of the process that led to the 'open data' needs to be as comprehensive as possible and the FAIR metadata of the restricted set should still be open access, including the accessibility protocol being well explained (licenses, procedures to follow if access to the restricted data is to be requested, etc.). However, data publishing may be many orders of magnitude more expensive than publishing narrative. A small example: One modern instrument (for instance a cryo-electron microscope) easily produces around 1 terabyte (TB) per day. Thus, a typical high-end instrument research group that uses the available instrument time efficiently produces 100-300 TB per year. Typically, projects take 1 - 3 years, resulting in a requirement of approximately 200 - 600 TB of storage for active projects. Let's assume that the funder of the research requires that after completion, the data must be archived for at least 10 years. In that case, a data intensive research group may have to archive 2 - 6 petabytes of raw data over 10 years. Current (academic) prices are in the range of 50 Euro/TB/year + 180 Euro/TB/year for backup and for archiving. So, storing 1000 TB in a proper way might add up to well over 100,000 Euro/year, just for archiving. This means that a data steward needs to budget very carefully for data storage, archiving and backup.

A very important decision is: what part of the data to publish, where, and how. Storing metadata or annotations in a FAIR format and in an HPR environment is almost trivial from a cost perspective, and may be enough to make the actual data optimally reused.

DO

- Include resources in your study budget for the FAIR publishing of your data and/or metadata, regardless of whether they can be fully open access.

- Be aware that publishing large datasets may actually be much more expensive than publishing the accompanying article.

- Work closely with experienced FAIR data colleagues and data publishers inside or outside your group to make sure that you indeed make the best (and affordable) technological, format, and terminology choices that can be considered FAIR.

- Choose a trusted data repository, preferably certified, to deposit the data once they are in FAIR publishable format.

- Choose an appropriate license for the the metadata and the data and include a machine-readable identifier for the data license in your metadata.

- Make sure your data are citable.

- Consider raw, processed, and metadata separately for publication choices.

- Separate these data categories but ensure a resolvable permalink between them.

DON'T

- Publish (supplementary) data in PDF or any other non-machine-readable format, rendering them elusive for most text-mining programs.

- Assume that publishing the data coming from your experiment requires a budget that is in the margin, or even in the range of the APC of the articles you publish.

- Publish any data without proper reference to the workflows and other research objects that were used to generate and analyse them.

- Publish without a serious check on price, trust and quality of the provider.

Resources: `http://dmp.fairdata.solutions/resources/mjf`

8.3 LEGAL ISSUES

What's up?

Data are very different from text. Text is usually seen as a *creative arte-fact* and is therefore 'by definition' subject to copyright. There is an enormous amount of recent literature about the use and abuse of copy-right in the current publishing world and how this jeopardises science. It is worthwhile to restate and discuss all the arguments here. How-ever, it is clear that 'facts' are not necessarily subject to copyright, and the question is whether 'collection of facts' are. Now, we try to avoid this unfruitful discussion altogether because we can introduce the no-tion that the 'data owner' (the collector or the operator/controller) has the 'right' to make the data FAIR (accessible under well-defined conditions). So the basic rule is: Never transfer rights on anything to a publisher, but keep control of the access rights to your data (and your articles, by the way, but that is less important for data stew-ardship). What is more important is: Are you really the legal 'owner' of the data you collected for your research? This may not always be the case. There might be an intrinsic ownership of the data. For in-stance, personal health data do not necessarily legally belong to the data provider, but at least the right to control access to the data lies with the citizen participating in a study unless explicitly transferred to the researcher. So in brief: Do you have the authority to determine the level of accessibility of the data in question? One specific question that will play a role here is that in case you use OPEDAS in your research and they have significantly contributed to the creation of your newly interpreted data and claims, you should first of all cite those properly (if possible), and secondly, you should make sure that reusing part of these data (even if you publish basic triples in a broader graph) did have a license that allows you to re-publish and re-distribute them. The latter is different from the consent in that you can use them for your research and internal analysis.

DO

- Make it absolutely clear that the rightful ownership situation of the data you have collected is beyond doubt.

- Consult an ELSI or other legal expert in case there is any doubt about such issues.

- Make sure all consent rules are followed, not only on newly generated data, but include OPEDAS, and include required citations.

DON'T

- Assume that data (also OPEDAS) with no license are open and reusable by default. They will be excluded by many (especially commercial data services and companies).

- Forget to separate consent to use and right to re-publish or re-distribute or to address both in concert.

- Republish OPEDAS (elements) without clear allowance to do so (based on license levels) or consent from the OPEDAS owner.

8.3.1 Where to publish?

What's up?

Data publishing is not the same as data archiving. The term 'publishing' has its roots in the notion of making assets (traditionally text or images) 'public'. No matter how much this term has undergone semantic drift in the current scholarly 'publishing' practice, we wish to stick to the original notion. So, data archived in your local repository are not automatically 'published'. There may be a very good reason to keep them just 'archived' for internal reuse, obviously with all the backup and safety issues discussed earlier, but here we assume that you really want to publish your data for external reuse. These costs should be eligible for the funder of your research and are comparable in nature to the article processing charge (APC) for open access articles. However, unless explicitly mentioned in a contract it is usually not the responsibility of the creator of the data to keep the data in an HPR environment for many years for others to reuse. Reuse of OPEDAS is costly, and part and parcel of proper budgeting for modern, data-intensive research projects. So the actual reuse of OPEDAS in HPR formats should be eligible research costs in future grants. Open data are not free in terms of *gratis* to reuse. This includes data, and data infrastructure resources that are very intensively used. These should be recognized at some point as *core infrastructure* and partly funded as a common good and/or from reuse fees. In the total offering of open access publishers, obviously reliable reuse of the article, in principle, for an indefinite time, is included. Storing huge datasets, however, will be significantly more expensive (at least initially) than storing text. So, long term-preservation costs need to be taken into account. There is also a major difference between publishing (FAIR) data in a basic trusted repository where both people and machines will be able to find them, knowing under which conditions to access them (the F + A of FAIR) and offering your data in fully interoperable formats, in a high-performance reusability (HPR) environment. Here, we warn you that HPR provision of your data to others is typically at least an order of magnitude more costly in terms of hardware and accompanying service and support than just 'publishing them' in a format that makes them principally FAIR. If your data are published, found, and reused by others as OPEDAS, the new user will have to bear the costs to actually include the data in a re-analytics environment. So, publishing data is a different decision from maintaining (yourself) an HPR environment for them. Writing massive data to tape and storing their rich and FAIR metadata in the cloud will usually still be considered good data stewardship. The tapes can be found, accessed, and the data retrieved in interoperable and thus reusable format, may it be at significant costs, but these should be borne by the reuser. Your choice of a repository is extremely important. Not only may the costs differ significantly between the many emerging professional data publishers, but matters of trust, sustainability and support issues are minimally as important. In very general terms, professorware-based local repositories are about the worst place to publish your data. In addition, there may soon be as many *data sharks* in the market soon as there are text sharks (predatory publishers) today. So, be aware of the strengths and weaknesses of various data publication options, and don't take any beautiful and promising website at face value. The general attitude you may best adopt as a good data steward in open science is that money is not made on the data (as an asset with paid access) but on services *around* the data, including HPR services, smart analytics, etc.

DO

- Publish data either internally or externally, but in all cases make sure you use reliable and sustainable repositories.

- If published externally, make sure to use one of the certified trusted data depositories and trusted publishing groups.

- Publish your (FAIR) metadata about the dataset in a public repository, regardless of where the data themselves are stored.

- Distinguish between inert publishing and offering data in high-performance reusability environments, and budget for both separately.

- Make it extremely clear, as part of the FAIR metadata, under which conditions the data can and may be reused. This goes way beyond a license, and may, for instance, tell the user (frequently a machine), where to find the raw data and which steps are necessary to reload them in an HPR.

DON'T

- Confuse long term archiving with initial publishing.

- Confuse inert publishing with offering an HPR environment.

- Think that keeping data internal (even if for very good reasons, such as patient privacy) relieves you from the good data stewardship practice of publishing rich and FAIR metadata about the internal dataset in a public and findable place, even if those metadata tell the potential reuser that reuse will be highly restricted.

- Think too easily that your data have not been produced with public funding, and therefore, you are not morally obliged to at least expose the metadata to the rest of the community.

- Think that when you work in a private company all this does not apply to you. As long as private companies can deduct research activities as eligible tax exempt costs, the community is co-financing your research.

Resources: `http://dmp.fairdata.solutions/resources/jbz`

8.4 WHAT TECHNICAL ISSUES ARE ASSOCIATED WITH HPR?

What's up?

Offering your data in an HPR environment is much more costly than just publishing them. However, there might be several (and rapidly accumulating) reasons to choose an HPR environment anyway. First, in case you are internally rewarded for the reuse of your data (unfortunately, only a fledgling practice to date), there may be great benefits in choosing an HPR, because the chances that others will actually reuse and cite your data will naturally be higher. Second, the participation in a FAIR HPR environment may give you extremely valuable feedback on your data over time, not only through reuse *per se*, but, for instance, also through annotation (by others) and by progressing insight created through other people's research that may reveal new patterns or associations in your data, which may lead to further insights, claims and (possibly joint) publications. Therefore, considering the much higher costs of offering the data yourself in an HPR or submitting them to one of the emerging HPR environments is a worthwhile exercise. In many cases, private providers will be eager to include your data in their HPR environments (such as high-performance analytics graphs and visualisations). This may sometimes only pertain to part of your data, but it could expose your data in multiple HPR environments without your bearing the costs of the provision in those environments. The HPR providers make a living by reformatting, analysing, linking, and integrating your data with OPEDAS, and, although the data as such are open and freely delivered to the HPR provider, an entire market in value-added services around data is rapidly developing while you read.

In case you decide with the team that running your own HPR is not within your reach for technical, expertise, support, or financial reasons, you may still consider offering your data formally for reuse in an existing HPR environment (open access or restricted and/or commercial). In other words, consider the 'value' of your data up to the point where they may be an asset for commercial HPR providers in the cloud that may help you in publication (for free) in their desired format, or may even pay you for your data. Also consider that funders will usually ask you to publish your data FAIR-compliant, but that offering them in an HPR environment is an added-value service that is normally not required.

DO

- Carefully consider the pros and cons of HPR provision of your published metadata and/or data.

- Clearly distinguish between 'local' HPR environments (very costly) and joining an existing HPR provider, where you let them offer your data in the desired format.

- Discuss the 'value' of your data in the team, with your supervisors, and make sure that due credit is given once your data actually does get reused and cited.

- Carefully check what the actual requirements of your funder are for publishing, long term preservation, and/or HPR options.

DON'T

- Lightly assume that just publishing your data (albeit FAIR) in an 'inert' repository is always just enough to claim good data stewardship. It certainly is from a minimalist standpoint, but it may severely limit your own chances to 'dig more gold from your data' later.

- Think that offering your data in an HPR environment will jeopardize or lower your competitive advantage. Especially for big datasets, there is a good chance that the participation in a broader set of frequently meta-analysed data will reveal new patterns (to others, in which case you should be cited, but also to you).

- Mix up publishing costs and HPR related costs in your data stewardship plan.

Resources: `http://dmp.fairdata.solutions/resources/xke`

8.4.1 What service will be offered around your data?

What's up?

In case you decide on more than 'inert publishing' or archiving in your *own institute*, you have to carefully manage expectations of your own research team and others on the level of services you provide on the data once it is published. Even if you do not opt for an internal HPR service, there may be significant costs involved if you ask an external provider to serve up your data in an HPR environment. This is a relatively new field, so there might be regular breakthroughs and incremental changes, and it may be wise to regularly check the emerging possibilities to put an 'active' copy of your data in an HPR environment.

DO

- Carefully consider the services you would like to attach to your data, in both cases, whether you decide on internal or external hosting.

- Make sure that if you decide on 'active' data hosting, you have sufficient resources to support the HPR environment you choose.

- Ensure that the analytics workflows, and other data-related services that you may offer, answer to FAIR principles themselves.

- Include metadata on the 'status' of the data in terms of their 'immediate' reusability.

- Include information about the possibility that your data can be accessed in 'inert' as well as in HPR environments, and where.

DON'T

- Call an inert data repository an HPR environment.

- Promise people that they can 'reuse' your data without specifying the conditions, and including the 'state' in which the data are offered (archived or HPR).

- Expect your high quality data to be intensively reused and cited if they are in an elusive local repository, highly restricted, or in a lousy state of interoperability.

Resources: `http://dmp.fairdata.solutions/resources/ivg`

8.4.2   Submit to an existing database?

What's up?

In some cases it might be wise to add your data to existing collections, such as international archives. More and more funding organisations and publishers will actually request that. This is in itself a reasonably straightforward process, as such public archives are usually maintained by dedicated institutes or consortia, and these have rules, regulations, requested formats, and, in many cases, specific instructions on how to upload and retrieve your data. Please note that some environments with be 'inert' non-interoperable archives (or only meant to serve human reuse) and some may qualify as machine-friendly HPR environments. Publishing in the latter category may pose some quite different challenges for you as a data creator (see also earlier considerations on updates and versioning of growing or changing datasets or resources). In many cases you may need to budget for professional assistance during the publishing process.

A more detailed consideration may be whether you consider your data 'reference' data that may be offered for curation, and including the addition to core data resources such as UniProt. In that case the procedures may be quite different and an active interaction with the data resource's custodians is in many cases the best way to proceed. Currently, many of these resources have to painstakingly recover the data they want to use in their curated and value-added resource by *ocular extraction* (reading) or by text and data mining, both of which are cumbersome and error prone. The *direct addition* of your data in the proper, unambiguous format to these core resources is part of good data stewardship practice, and will greatly enhance their reuse, and citability.

DO

- Always consider the 'potential reference value' of your data (for instance, can my new findings enrich reference data sources such as Chembl, UniProt, or EarthCube).

- Submit (parts of) your data to the appropriate 'archives', but also provide selected parts of your data to reference core databases whenever appropriate.

- Use the correct formats and standards required by those data resources and, if needed, contact them.

- Select the nest databases not only by quality, but also appropriateness for the type of data.

DON'T

- Just publish your data as 'supplementary files' with your article(s) and assume that professional data custodians of archives, HPR environments and core reference databases will find and use them independently.

- Upload data to such resources without proper metadata and provenance attached.

- Bother professional custodians with data that are clearly not suited to be reused.

Resources: `http://dmp.fairdata.solutions/resources/bvq`

8.4.3   Will you run your own access Web service for data?

What's up?

Although we recommend using trusted external HPR services wherever possible, there may be reasons to run your own access service: For instance, when data you generated cannot legally or technically leave your institutional firewall. In that case, you can only offer them for reuse by third parties by providing the data in a reusable format and allow 'workflows ' (here used in the broadest possible sense) to visit the data. In that case, you need to make sure that you have enough local compute power to run the workflows, and that you can properly handle controlled access, authentication, authorisation, encryption, logging, and monitoring of what goes in and out, as an internal service. This is far from trivial and will pose significant challenges to your budget, infrastructure, and internal expertise. In case it is unavoidable to keep data 'inside your firewall', there are serious management issues to be addressed, like the training of support personnel and the acquisition (hardly ever the development) of standard software, hardware, and services to allow all the data management access and processing steps mentioned above. The good news is that more standardized solutions for visiting workflows are under development.

You also need to be fully aware of the various levels of guaranteed quality of service you want to offer, such as 'up time': only at written request and *ad hoc*, academic best practice, 24/7, etc., and the budgetary consequences of such choices. For instance, offering '24/7' may be an order of magnitude more complicated and costly than academic best practice access (the data is up if you are lucky; don't blame me if there is a problem that jeopardised your three-day running workflow because my component was 'down'). The components you need to consider include (not exhaustive): a processing service, including an authentication and authorisation infrastructure (AAI), a download service, a hosting service, system, hardware, software, financial administration, and user logging. You also need to consider long-term sustainability of the service and who keeps data access running, and the installation, staffing, and training for a help desk for data access and interpretation.

DO

- Carefully consider the reasons why data are not supposed to leave your fire-walled institutional environment.

- Keep the amount of data offered for reuse where this is indeed unavoidable to the bare minimum, as they may cost you a lot.

- As much as possible, use standardised, reliable public or private software, hardware and software providers in order to enable the services needed to make data available under active reuse conditions.

- Budget very carefully for the considerable costs involved.

DON'T

- Develop any in-house tools, algorithms, or software needed to offer 'active' data under controlled conditions unless absolutely unavoidable.

- Allow that even if the data are of the most restricted category, the metadata is still made available in the most open format that your restrictions allow, and include in the metadata, rich information about the legal and technical needs and restrictions for the reuse of the actual data.

- Describe your data as 'reusable in HPR' if the management components are not reliable (i.e., self-made) and your data component may therefore cause much trouble in pipelines or workflows of others because they did consider your Web service as a reliable component of their pipeline/workflow.

8.4.4   How and where will you be archiving/cataloguing?

What's up?

It is obvious that only the 'active' version of any dataset that is offered for direct reuse ever has to be in an HPR environment of any sort, and may be replicated in several of them for different purposes. The backup(s) of the dataset may be in a totally different format and accessibility status, for instance on tape. Still, you need to be aware that in case the 'HPR version(s)' of the data get lost or corrupted, you are able to regenerate the HPR version of the data at short notice and without errors. It is crucially important that the location and the access procedures of the backup data that are archived are clearly stated in your institutional catalogue and not just understandable by you as the data steward. A good data steward will always consider the possibility that others will need to use any data or service (including the remounting of archived data) in situations where the original data steward is not available for consultation or help.

DO

- Archive the backups of the data you provide for active reuse in the most stable and sustainable (also the cheapest) way and format possible.

- Include the metadata and provenance of the archived (backup) data in the metadata files of the active data version.

- Provide clear instructions for anyone in the institute on how to find, process and remount the archived data into the format and the environments from which the original may have been lost or corrupted.

- Regularly check the 'mounted' data in HPR for integrity and only regenerate them when needed with full records and provenance of when, why, and how the data were regenerated and remounted.

DON'T

- Assume that it is sufficient that you yourself know where the archives are and how they can be accessed.

- Consider a 'backup' a simple copy of the entire HPR environment (considering that there is an order of magnitude higher effort and cost associated with the latter).

- Just archive 'everything', but carefully consider what has to be archived as even archiving is very expensive if you enter the tera- and petabyte range.

Resources: `http://dmp.fairdata.solutions/resources/fxe`

8.5 WILL YOU STILL PUBLISH IF THE RESULTS ARE NEGATIVE?

One of the most fundamental flaws of the current scholarly communication practice via journals and articles is the near-impossible publication of negative results in peer-reviewed journals. The current upcoming post-publication peer review approaches may start to alleviate this problem, but it should be clear to any data steward that 'negative results' are not at all to be excluded from the publication practice. On the contrary, they should be a key part of the FAIR research object ecosystem. The fact that others have 'tried a logical but apparently dead end', or they have contested earlier claims, or report that they are unable to reproduce data creation, results, or correlations published earlier, can critically improve the efficiency of open science, if only by avoiding repetition of approaches leading to negative or confusing results.

We recommend that you use suitable (post-publication review) and machine-readable (such as nanopublication) publication formats to maximally ensure that negative results are properly published, with a supplementary paper describing the original study and the reasons why the data are considered negative, or meaningless, in the sense that they do not support the hypothesis tested, or they have shown, for instance, that the experimental setting was not adequate to achieve the proper outcome, and the reasons why. This category could also include data that, for instance, show why a misinterpretation was made (even in hindsight) or why a certain setup for a cell culture did not work, or that the wrong cell line or buffer was used.

8.5.1 Data as a publishable unit?

What's up?

Even if it is impossible to publish your 'negative results' in a classical journal, the data supporting your conclusions (even if an experiment is not reproducible or you claim that earlier results are wrong) can always be published in their own right, accompanied by minimal text to explain what the issue is. It is crucial that the data you publish this way is FAIR, as machines should also be aware of controversies, negative results and other 'data warnings'. Linking negative or controversial results to the original papers (DOIs) is important as it allows certain services to automatically annotate these older papers with 'second thought' remarks, even in PDF (see for instance UTOPIAdocs).[3]

DO

- Publish all solid, high quality data and results, regardless of whether they are considered 'positive' in the sense of confirming your hypothesis. The knowledge that 'something is not true or does not work' is also a relevant new insight.

- Consider that many data may be more hypothesis-free than you might think, even if you generated them with a particular hypothesis in mind.

- Publish negative or controversial assertions as machine-readable nanopublications and push them for peer review.

---

[3]https://www.youtube.com/watch?v=1M26DRlZwSM

DON'T

- Ignore negative results unless you can explain why the results do not represent valid new associations or insights, for instance, because review of the experimental set-up or the study reveals fundamental flaws. Even in that case the reasons for the flawed set-up may be worth publishing because it may prevent others from making the same mistake.

- Try to publish negative results in classical narrative articles/journals, especially when you try this in pre-peer-review journals, which will only yield frustration.

- Give up too easily when negative results are difficult to publish, as they may appear to be a major contribution to science and heavily cited.

Resources: `http://dmp.fairdata.solutions/resources/xvp`

8.5.2   Will you publish a narrative?

What's up?

Disclaimer: The arguments put forward in this section are mostly relevant for hard-core scientific communication, although any text (including this book) should benefit from taking many of the considerations into account. As said before, narrative articles (also called 'papers') will likely continue to exist for the foreseeable future, as they still have their value, mostly for the communication of human interpretation of results. Computers cannot easily cover rhetorical strings and arguments, and therefore a narrative (although inherently ambiguous and thus not FAIR) is needed to explain what was discerned in the data, how the data were generated, and what formed the basis for the conclusions. Part of all this, including methods etc., can increasingly also be captured in machine-actionable format, and therefore become part of the machine-actionable FAIR ecosystem. But 'supplementary narrative text' with your data will always be of additional value and obviously, review articles, describing a whole field, will always play a role as long as human minds are part of the scientific process. However, when you write a narrative, in whichever component of what you want to publish, take care *never* to introduce ambiguous symbols and formulations in the text. Narrative with synonyms and beautiful style figures may be entertaining for people to read, but these usually do not at all serve well-defined and straightforward scientific reasoning, but rather blur it. More importantly, they make the text even more of a nightmare for machine-reading, text- and data-mining than they already are by their very lingual nature. Unexplained acronyms (*people will know what I mean*) with multiple meanings (homonyms), alliterations, nested sentences, and sentences of the style *concept 1, 2, 3, 4 and 5 are all related to concept 6 in table 5* are *absolute malpractice* in scientific text. In fact, a good exercise to internally check whether any narrative you add to your data is reasonably unambiguous, is to first represent as much as you can of the reasoning in a machine-readable format and formulate text based on this template. You should publish those machine-readable assertions along with the actual text. This means that if you use terms such as 'intellectual disability' and 'mental retardation' interchangeably in the same text, or the same symbol for a gene as well as its corresponding protein (which are two distinct concepts), machines (and people alike) can at least check the corresponding unique identifiers used in the machine-friendly version in order to 'understand' what you actually mean. This exercise also forces you to structure your reasoning as precisely and logically as possible, so it will also improve the quality of your text. Also, refer to equipment, reagents, and methods with unique identifiers linked to the term, and avoid using unspecific terms and jargon. Sentences that really cannot be represented in a machine-readable format, even after you have tried, probably represent exactly that (crucial) portion of your narrative, that is needed to explain to other people how, why, and on what basis you argue for your conclusions and claims.

DO

- Take great effort to make any narrative that you have to produce as straightforward and unambiguous a text as possible, to make it more readable for both machines and people. Scientific text is not for leisure reading and entertainment, but should serve unambiguous scientific and scholarly communication wherever possible.

- Only use a narrative if you address people, and not when your user is likely to be a machine.

- Reduce the amount of narrative you produce to the minimum needed to communicate your methods, reasoning, and rhetoric. We are bombarded with text to the extent that we need to read 70 hours a day to keep up with the 'literature' that is out there already.

DON'T

- Use jargon, undefined acronyms or synonyms in your narrative, unless really unavoidable.

- Produce one line more of text than is needed to effectively communicate your point.

- Use free text wherever machine readable and actionable formats can also do the job. There will still be enough text left, don't worry.

- Add references to data in the text which are not machine resolvable.

Resources: `http://dmp.fairdata.solutions/resources/igj`

Bibliography

[Barnes et al., 2009] Barnes, M. R., Harland, L., Foord, S. M., Hall, M. D., Dix, I., Thomas, S., Williams-Jones, B. I., and Brouwer, C. R. (2009). Lowering industry firewalls: pre-competitive informatics initiatives in drug discovery. 8(9):701–708.

[Bechhofer et al., 2010] Bechhofer, S., Roure, D. D., Gamble, M., Goble, C., and Buchan, I. (2010). Research objects: Towards exchange and reuse of digital knowledge. (713).

[Fehrmann et al., 2015] Fehrmann, R. S. N., Karjalainen, J. M., Krajewska, M., Westra, H.-J., Maloney, D., Simeonov, A., Pers, T. H., Hirschhorn, J. N., Jansen, R. C., Schultes, E. A., van Haagen, H. H. H. B. M., de Vries, E. G. E., te Meerman, G. J., Wijmenga, C., van Vugt, M. A. T. M., and Franke, L. (2015). Gene expression analysis identifies global gene dosage sensitivity in cancer. 47(2):115–125.

[Freedman et al., 2015] Freedman, L. P., Cockburn, I. M., and Simcoe, T. S. (2015). The economics of reproducibility in preclinical research. 13(6):e1002165.

[Groth et al., 2010] Groth, P., Gibson, A., and Velterop, J. (2010). The anatomy of a nanopublication. 30(1):51–56.

[Ioannidis, 2005] Ioannidis, J. P. A. (2005). Why most published research findings are false. 2(8):e124.

[Jimeno Yepes and Verspoor, 2014] Jimeno Yepes, A. and Verspoor, K. (2014). Literature mining of genetic variants for curation: quantifying the importance of supplementary material. 2014.

[Longo and Drazen, 2016] Longo, D. L. and Drazen, J. M. (2016). Data sharing. 374(3):276–277.

[Mayer and Rauber, 2015] Mayer, R. and Rauber, A. (2015). A quantitative study on the re-executability of publicly shared scientific workflows. In *2015 IEEE 11th International Conference on e-Science*, pages 312–321.

[Mons, 2005] Mons, B. (2005). Which gene did you mean? 6:142.

[Mons et al., 2017] Mons, B., Neylon, C., Velterop, J., Dumontier, M., Santos, d. S., Bonino, L. O., and Wilkinson, M. D. (2017). Cloudy, increasingly FAIR; revisiting the FAIR data guiding principles for the european open science cloud. 37(1):49–56.

[Mons et al., 2011] Mons, B., van Haagen, H., Chichester, C., Hoen, P.-B. t., den Dunnen, J. T., van Ommen, G., van Mulligen, E., Singh, B., Hooft, R., Roos, M., Hammond, J., Kiesel, B., Giardine, B., Velterop, J., Groth, P., and Schultes, E. (2011). The value of data. 43(4):281–283.

[Mons and Velterop, 2009] Mons, B. and Velterop, J. (2009). Nanopublication in the e-science era. In *Workshop on Semantic Web Applications in Scientific Discourse (SWASD 2009)*, pages 14–15.

[Musen et al., 2015] Musen, M. A., Bean, C. A., Cheung, K.-H., Dumontier, M., Durante, K. A., Gevaert, O., Gonzalez-Beltran, A., Khatri, P., Kleinstein, S. H., O'Connor, M. J., Pouliot, Y., Rocca-Serra, P., Sansone, S.-A., Wiser, J. A., and CEDAR team (2015). The center for expanded data annotation and retrieval. 22(6):1148–1152.

[Read et al., 2015] Read, K. B., Sheehan, J. R., Huerta, M. F., Knecht, L. S., Mork, J. G., Humphreys, B. L., and Group, N. B. D. A. (2015). Sizing the problem of improving discovery and access to NIH-funded data: A preliminary study. 10(7):e0132735.

[Rebholz-Schuhmann et al., 2012] Rebholz-Schuhmann, D., Oellrich, A., and Hoehndorf, R. (2012). Text-mining solutions for biomedical research: enabling integrative biology. 13(12):829–839.

[Rodríguez-Iglesias et al., 2016] Rodríguez-Iglesias, A., Rodríguez-González, A., Irvine, A. G., Sesma, A., Urban, M., Hammond-Kosack, K. E., and Wilkinson, M. D. (2016). Publishing FAIR data: An exemplar methodology utilizing PHI-base. 7.

[Sharing, 2016] Sharing, T. I. C. o. I. f. F. i. T. D. (2016). Toward fairness in data sharing. 375(5):405–407.

[The FANTOM Consortium and the RIKEN PMI and CLST (dgt), 2014] The FANTOM Consortium and the RIKEN PMI and CLST (dgt) (2014). A promoter-level mammalian expression atlas. 507(7493):462–470.

[Wilkinson et al., 2016] Wilkinson, M. D., Dumontier, M., Aalbersberg, I. J., Appleton, G., Axton, M., Baak, A., Blomberg, N., Boiten, J.-W., da Silva Santos, L. B., Bourne, P. E., Bouwman, J., Brookes, A. J., Clark, T., Crosas, M., Dillo, I., Dumon, O., Edmunds, S., Evelo, C. T., Finkers, R., Gonzalez-Beltran, A., Gray, A. J., Groth, P., Goble, C., Grethe, J. S., Heringa, J., 't Hoen, P. A., Hooft, R., Kuhn, T., Kok, R., Kok, J., Lusher, S. J., Martone, M. E., Mons, A., Packer, A. L., Persson, B., Rocca-Serra, P., Roos, M., van Schaik, R., Sansone, S.-A., Schultes, E., Sengstag, T., Slater, T., Strawn, G., Swertz, M. A., Thompson, M., van der Lei, J., van Mulligen, E., Velterop, J., Waagmeester, A., Wittenburg, P., Wolstencroft, K., Zhao, J., and Mons, B. (2016). The FAIR guiding principles for scientific data management and stewardship. 3:160018.

Index